Lecture Notes in Statistics   160

Edited by P.Bickel, P. Diggle, S. Feinberg, K. Krickeberg,
I. Olkin, N. Wermuth, and S. Zerger

Springer
*New York
Berlin
Heidelberg
Barcelona
Hong Kong
London
Milan
Paris
Singapore
Tokyo*

Tomasz Rychlik

Projecting Statistical Functionals

 Springer

Tomasz Rychlik
Institute of Mathematics
Polish Academy of Sciences
Chopina 12
87100 Toruń
Poland
trychlik@impan.gov.pl

Library of Congress Cataloging-in-Publication Data
Rychlik, Tomasz.
    Projecting statistical functionals / Tomasz Rychlik
      p.   cm. — (Lecture notes in statistics; 160)
    Includes bibliographical references and index.
    ISBN 0-387-95239-X (softcover: alk. paper)
    1. Statistical functionals. 2. Hilbert space. I. Title. II. Lecture notes in
    statistics (Springer-Verlag); v. 160.
QA273.6 .R93 2001
519.2'4—dc21   00-069240

Printed on acid-free paper.

© 2001 Springer-Verlag New York, Inc.
All rights reserved. This work may not be translated or copied in whole or in part without the written permission of the publisher (Springer-Verlag New York, Inc., 175 Fifth Avenue, New York, NY 10010, USA), except for brief excerpts in connection with reviews or scholarly analysis. Use in connection with any form of information storage and retrieval, electronic adaptation, computer software, or by similar or dissimilar methodology now known or hereafter developed is forbidden. The use of general descriptive names, trade names, trademarks, etc., in this publication, even if the former are not especially identified, is not to be taken as a sign that such names, as understood by the Trade Marks and Merchandise Marks Act, may be accordingly used freely by anyone.

Camera-ready copy provided by the author.
Printed and bound by Sheridan Books, Inc., Ann Arbor, MI.
Printed in the United States of America.

9 8 7 6 5 4 3 2 1

ISBN 0-387-95239-X   SPIN 10793590

Springer-Verlag   New York Berlin Heidelberg
A member of BertelsmannSpringer Science+Business Media GmbH

Preface

About 10 years ago I began studying evaluations of distributions of order statistics from samples with general dependence structure. Analyzing in [78] deterministic inequalities for arbitrary linear combinations of order statistics expressed in terms of sample moments, I observed that we obtain the optimal bounds once we replace the vectors of original coefficients of the linear combinations by the respective Euclidean norm projections onto the convex cone of vectors with nondecreasing coordinates. I further verified that various optimal evaluations of order and record statistics, derived earlier by use of diverse techniques, may be expressed by means of projections. In Gajek and Rychlik [32], we formulated for the first time an idea of applying projections onto convex cones for determining accurate moment bounds on the expectations of order statistics. Also for the first time, we presented such evaluations for nonparametric families of distributions different from families of arbitrary, symmetric, and nonnegative distributions. We realized that this approach makes it possible to evaluate various functionals of great importance in applied probability and statistics in different restricted families of distributions.

The purpose of this monograph is to present the method of using projections of elements of functional Hilbert spaces onto convex cones for establishing optimal mean-variance bounds of statistical functionals, and its wide range of applications. This is intended for students, researchers, and practitioners in probability, statistics, and reliability. Intended as a reference book it could also be used as a textbook for a specialized course in the subject area. Numerous open problems are formulated in the text. I hope that they stimulate some readers to undertake research in this direction.

The prerequisites are upper-level undergraduate courses in probability theory, mathematical statistics, and functional analysis, including an elementary theory of Hilbert spaces. Certainly, some knowledge of nonparametric statistical inference and reliability theory would be beneficial. It is important to become acquainted with the content of Chapter 2 before reading the following ones. Chapters 3 through 6 can be studied independently, with some exceptions. Some results of Sections 5.3, 5.4, and 5.6 are deduced from lemmas contained in Sections 3.2, 3.3, and 4.5, respectively. The results of Chapter 7 are based on applications of second moment bounds presented in Sections 4.1, 4.2, 6.2, and 6.3.

This work was done with the support of the Polish State Committee for Scientific Research (KBN) Grant 2 P03A 014 13. The idea of writing the book was suggested by Lesław Gajek. Besides working together on specific problems, we carried on comprehensive discussions that enabled us to crystallize original vague ideas into a definite plan. As he undertook other important obligations, he could not participate in completing the project. Some results were obtained in cooperation with Andrzej Okolewski whose contribution is gratefully acknowledged. Significant parts of the content of this book were presented and discussed during regular Thursday seminars on applied probability held in the Institute of Mathematics of Polish Academy of Sciences in Warsaw. Critical remarks and comments of the participants had a strong positive impact on the final outcome. The assistance of several people made it possible to improve the presentation. Many helpful suggestions were provided by John Kimmel, the Executive Editor of the Statistics Division at Springer-Verlag, and one of the reviewers. The copyeditor corrected a number of linguistic and style mistakes. Jan K. Kowalski helped to prepare the final LATEXversion of the manuscript in camera-ready form. It is a pleasure to offer my sincere thanks to them.

Tomasz Rychlik
January, 2001

Contents

Preface	v
List of Tables	ix

1 Introduction and Notation   1
   1.1 Introduction 1
   1.2 Notation 5

2 Basic Notions   11
   2.1 Elements of Hilbert Space Theory 11
   2.2 Statistical Linear Functionals 16
   2.3 Restricted Families of Distributions 25

3 Quantiles   33
   3.1 General and Symmetric Distributions 33
   3.2 Distributions with Monotone Density and Failure Rate . . . 36
   3.3 Distributions with Monotone Density and Failure Rate on the Average 44
   3.4 Symmetric Unimodal Distributions 50
   3.5 Open Problems 54

4 Order Statistics of Independent Samples   55
   4.1 General and Symmetric Distributions 56
   4.2 Life Distributions with Decreasing Density and Failure Rate   60

	4.3 Distributions with Monotone Density and Failure Rate on the Average	69
	4.4 Symmetric Unimodal Distributions	79
	4.5 Bias of Quantile Estimates	82
	4.6 Open Problems	91

5 Order Statistics of Dependent Observations — 95

5.1 Dependent Observations with Given Marginal Distribution — 96
5.2 General and Symmetric Distributions — 100
5.3 Distributions with Monotone Density and Failure Rate — 105
5.4 Distributions with Monotone Density and Failure Rate on the Average — 114
5.5 Symmetric Unimodal and U-Shaped Distributions — 115
5.6 Bias of Quantile Estimates — 118
5.7 Extreme Effect of Dependence — 121
5.8 Open Problems — 127

6 Records and kth Records — 131

6.1 Dependent Identically Distributed Observations — 131
6.2 General and Symmetric Distributions — 133
6.3 Life Distributions with Decreasing Density and Failure Rate — 136
6.4 Increments of Records — 140
6.5 Open Problems — 143

7 Predictions of Order and Record Statistics — 145

7.1 General Distributions — 146
7.2 Distributions with Decreasing Density and Failure Rate — 149
7.3 Open Problems — 154

8 Further Research Directions — 157

References — 163

Author Index — 171

Subject Index — 173

List of Tables

3.1 Sharp uniform mean-variance bounds on quantiles for various families of distributions. 53

4.1 Sharp uniform mean-variance bounds on expectations of order statistics from independent samples of size 20 for various families of distributions. 81
4.2 Sharp uniform variance bounds on upper bias deviations of estimators $X_{j:20}$ of pth quantiles, $p = j/20$, for various families of distributions (independent case). 92

5.1 Sharp uniform mean-variance bounds on expectations of order statistics from dependent samples of size 20 for various families of distributions. 117
5.2 Sharp uniform variance bounds on upper bias deviations of estimators $Y_{j:20}$ of pth quantiles, $p = j/20$, for various families of distributions (dependent case). 122

1
Introduction and Notation

1.1 Introduction

This work presents a method of using projections of functions onto convex cones in Hilbert spaces for determining sharp bounds on values of statistical functionals over general and restricted families of distributions, expressed in terms of moment parameters of the distributions. The method is based on representing the statistical functionals and families of distributions as fixed elements and convex cones, respectively, in a common real Hilbert space. Then the norm of the projection of the element onto the cone provides the optimal bound. The distribution for which the bound is attained is derived by a simple transformation of the projection.

The advantage of the projection method lies in providing definite answers for numerous simply stated but nontrivial problems of theoretical and practical importance. The method enables us to optimally evaluate various random objects in terms of the first two moments which are the most classical parameters describing the population. Natural restrictions on the random structure of observations are allowed. Although some results presented here were proven earlier by means of specific tools, our unified approach provides simpler proofs and indicates mutual relations among various evaluations. On the other hand, for numerous problems presented in the book, we do not see alternative ways of solving them without reference to projections. The idea of projections has been effectively exploited in various aspects of statistical inference, for instance in the least squares and minimum distance estimation. The majority of applications are based

2   1. Introduction and Notation

on projections onto finite-dimensional linear subspaces. The novelty of our approach consists in using solutions of projection problems onto convex cones in function spaces for precise evaluations of statistical functionals. We illustrate the general idea by an example.

EXAMPLE 1. Suppose that random variables $X_1, \ldots, X_n$ are independent identically distributed (i.i.d.) with a common distribution F. We are interested in evaluating the upper deviation of the expectation of the jth order statistic $E_F X_{j:n}$ from the population mean $\mu_F = E_F X_1$ in the standard deviation units $\sigma_F = (\text{Var}_F X_1)^{1/2}$. We can write

$$E_F X_{j:n} - \mu_F = \int_0^1 [F^{-1}(x) - \mu_F][f_{j:n}(x) - 1]\, dx, \qquad (1.1)$$

where

$$f_{j:n}(x) = n\binom{n-1}{j-1} x^{j-1}(1-x)^{n-j}$$

is the density function of the jth order statistic from the i.i.d. standard uniform sample. This may be interpreted as the inner product of the centered quantile function $F^{-1} - \mu_F$ with $f_{j:n} - 1$ in the Hilbert space $L^2([0,1), dx)$ of the square integrable functions on the unit interval. Applying the Schwarz inequality to (1.1), and noting that $\|F^{-1} - \mu_F\| = \sigma_F$, we obtain

$$E_F X_{j:n} - \mu_F \leq \|f_{j:n}(x) - 1\|\sigma_F. \qquad (1.2)$$

The equality is attained here if the arguments of the inner product are proportional; that is,

$$F^{-1}(x) - \mu_F = \alpha[f_{j:n}(x) - 1] \qquad (1.3)$$

for some $\alpha \geq 0$. All the possible centered quantile functions of distributions with a finite variance form the convex cone of nondecreasing functions integrating to 0 in $L^2([0,1), dx)$. The cone is further denoted by $\mathcal{C}^0$.

If $j = n$, then the right-hand side of (1.3) actually increases, and integrates to 0. Therefore bound (1.2) is then tight, and (1.3) enables us to determine a (power) distribution that attains the bound (see Gumbel [36], and Hartley and David [38]). Otherwise $f_{j:n} - 1$ is not nondecreasing, and (1.2) cannot be sharp.

The most natural idea is to replace $f_{j:n} - 1$ by the closest element of the family of centered quantile functions $\mathcal{C}^0$. We call it the projection of $f_{j:n} - 1$ onto $\mathcal{C}^0$ and denote it by $P^0(f_{j:n} - 1)$. The idea does work here and in many other problems considered in the monograph. Namely,

$$[E_F X_{j:n} - \mu_F]/\sigma_F \leq B^0(j,n) = \|P^0(f_{j:n} - 1)\| \qquad (1.4)$$

for all F with $\sigma_F^2 < \infty$, and we get the equality in (1.4) if

$$\frac{F^{-1}(x) - \mu_F}{\sigma_F} = \frac{P^0(f_{j:n} - 1)(x)}{\|P^0(f_{j:n} - 1)\|} \in \mathcal{C}^0, \qquad (1.5)$$

which defines distribution function F with the extreme normalized expectation of $X_{j:n}$. If $j = 1$, then $f_{j:n} - 1$ is decreasing, and the projection is constant
$$P^0(f_{j:n} - 1) = 0.$$
Hence we have $E_F X_{1:n} \leq \mu_F$ which is clear by relation $X_{1:n} \leq X_1$. If $1 < j < n$, then the increasing-decreasing function $f_{j:n} - 1$ is replaced by nondecreasing
$$P^0(f_{j:n} - 1)(x) = f_{j:n}(\min\{x, \alpha_*\}) - 1, \tag{1.6}$$
where α_* is a point of increase of $f_{j:n}$ determined by an equation (see Moriguti [58]). A detailed construction of (1.6) and verification of (1.4) and (1.5) is presented in Section 4.1 (see also Example 2 of Section 2.1). This is based on an old general method due to Moriguti based on convex minorants, and (1.6) is actually the derivative of the greatest convex minorant of the antiderivative of $f_{j:n} - 1$.

For the first time in the context of evaluating statistical functionals, however, the notion of projection appeared in Rychlik [78], where some deterministic bounds on linear combinations of order statistics in terms of sample mean and variance were established. The general idea of using projections for sharp evaluations of statistical functionals was formulated in Gajek and Rychlik [32]. The most recent review of results is presented in the expository paper by Rychlik [86]. In principle, we provide here an exposition of recent research: a significant number of papers we refer to have not been published yet, and some results have not been presented elsewhere. Since this work is entirely devoted to the projection method, we consistently omit discussing other approaches and many related results. Therefore references to results obtained by different methods are rare and laconic here. It is worth pointing out that considering Hilbert spaces and convex cones is essential for the method. For instance, the projections of linear statistical functionals in L^p-spaces with $p \neq 2$ do not provide analogous evaluations.

The structure of the book reflects the purpose of presenting bounds on various statistical functionals over different classes of distributions by means of the common method based on projections. Fundamental notions are introduced in Chapter 2. Some facts of general Hilbert space theory taht are applied in our method are collected in Section 2.1. Inner product formulae for some functionals with statistical interpretations are worked out in Section 2.2. In Section 2.3 special classes of distributions of practical importance in probability, mathematical statistics, and reliability are characterized by convex cones of respective quantile functions. Some partial orders of distributions that are useful in defining the classes are introduced there. Consecutive chapters are devoted to specific functionals, and bounds on the functionals over various classes are discussed in respective sections. Titles of sections refer to families of distributions with intuitive interpretations. In fact, more general results are presented. We often consider families of distributions related to a fixed general one with respect to partial orders defined in Section 2.3. Some results are specified for specially chosen extreme elements of the families, for instance for uniform and exponential distributions that generate some classes with natural intuitive properties.

The main results are presented in Chapters 3 to 7 in a unified way: each sharp bound on a fixed functional over a fixed class of distributions is explicitly written, together with a formula for the distribution function for which the bound is attained. Different bounds for various classes of distributions are also compared numerically. Special emphasis is laid on construction of projections, the essence of our method. Since in the problems we study, general methods for constructing projection functions are not known, various tools are needed for solving specific problems. Usually we first describe the shape of the projections up to several real parameters by means of geometric arguments, and then determine the optimal parameters analytically. Some bounds are expressed by complicated formulae that should be evaluated by use of subtle tools of numerical analysis. There is an apparent intentional contrast between precise formulation of results and informal justification. A detailed verification is provided for selected problems of bounds on quantiles and order statistics in Chapters 3 and 4, respectively. Presented here are the most typical geometric and analytic arguments used for determining projections of step and smooth functions. The other proofs are merely sketched and we refer the reader to the original papers for complete details. In particular, we omitted formal proofs of the results of Chapters 5 and 6, which are also substantially based on projecting step and smooth functions, respectively.

The results presented here are far from being complete. There are still many interesting open problems that can be formulated for statistical functionals and families of distributions described here. In the concluding sections of the chapters some open problems are stated. All of them can be reformulated as projection problems onto convex cones. They are more difficult than standard exercises and problems usually contained in monographs with the purpose of enabling the reader to gain a deeper understanding of the text. We hope that our problems stimulate readers to undertake research and deliver original and interesting evaluations of statistical functionals by means of projections. There are also other functionals and families of distributions that are not studied here and for which the projection method would work. Some of them are presented in Chapter 8, where possible further research directions are indicated. Extending the evaluation problems to these functionals and families generate new interesting questions. In order to solve them, one should possibly develop techniques and methods different from ones presented here.

1.2 Notation

We tried to set a unified notation for the whole work. A list of symbols is presented below. Only notation used locally is not included there. We generally preserved standard symbols. However, some other notation may look strange at the first glance. For instance, numerous convex cones of functions were considered in the text. They were denoted by $\mathcal{C}$ with various superscripts and subscripts whose meaning is explained in the list below (see also Section 2.3). The projection operator onto a given cone is written as P with the same upper and lower indices. Moreover, the indices appear in the notation of sharp bounds on functionals that are obtained by means of projections on specified convex cones. For instance, $B^0(j, n)$ calculated in Example 1, denotes the sharp mean-variance bound on the expectation of the jth order statistic of an independent identically distributed sample of size n with arbitrary marginal distribution that has a finite variance. The bound is obtained by means of projection P^0 of a properly chosen functional (see Section 2.2) onto the convex cone $\mathcal{C}^0$ of quantile functions of all distributions with finite variance, centered about the respective mean. Convex cones of quantile functions for families of distributions satisfying various restrictions and respective projection operators need more sophisticated notation. Nevertheless, we tried to introduce the symbols in a coherent manner and believe that the reader shall get used to them.

List of symbols

$(\mathcal{H}, (\cdot, \cdot))$	—	(real) Hilbert space with inner product $(\cdot, \cdot)$
$\|\cdot\|$	=	$(\cdot, \cdot)^{1/2}$ — norm in $(\mathcal{H}, (\cdot, \cdot))$
$L^2([a, d], w(x)\, dx)$	—	Hilbert space of functions $g : [a, d] \mapsto \Re$ satisfying $\int_a^d g^2(x) w(x)\, dx < \infty$ with inner product $(g, h) = \int_a^d g(x) h(x) w(x)\, dx$ for a positive weight function $w(x)$
$\alpha, \beta, \ldots$	—	scalars, parameters of functions
$\alpha_*, \beta_*, \ldots$	—	optimal scalars, parameters of projections
$g, h, \ldots$	—	functions, elements of Hilbert spaces
$G, H, \ldots$	—	antiderivatives of $g, h, \ldots$, respectively
$gH(x)$	=	$g(H(x))$ — composition of functions H and g
$\bar{H}(x)$	—	greatest convex minorant of $H(x)$
$\bar{h}(x)$	—	(right) derivative of $\bar{H}(x)$
$\mathbf{1}(x)$	—	constant function equal to 1
$\mathbf{1}_A(x)$	—	indicator of A ($= 1$ on A and 0 elsewhere)
x_+	=	$\max\{x, 0\}$ — positive part of a number
$\lfloor x \rfloor$	=	$\max\{k \leq x : k \text{ integer}\}$ — floor of a number

6   1. Introduction and Notation

$$
\begin{aligned}
T_h(\cdot) &= (h,\cdot) - \text{continuous linear functional on a Hilbert space} \\
T_h(\cdot)/\|\cdot\| &- \text{normalized linear functional} \\
F(x) &- \text{marginal distribution function of population} \\
f(x) &- \text{density function of } F \\
\lambda_F(x) &= f(x)/[1-F(x)] - \text{failure rate of } F \\
F^{-1}(x) &= \sup\{y: F(y) \le x\},\ 0 \le x < 1,\ - \text{quantile function of } F \\
F^{-1}(p) &- \text{quantile of order } p,\ 0 < p < 1 \\
F_{|y}(x) &= [F(x)-F(y)]/[1-F(y)],\ x \ge y,\ - \text{distribution function of } X \text{ under condition } X > y \\
F^0_{|y}(x) &= F_{|y}(x+y) = [F(x+y)-F(y)]/[1-F(y)],\ x \ge 0,\ - \text{distribution function of } X-y \text{ under condition } X > y \\
F^{|z}(x) &= F(x)/F(z),\ x \le z,\ - \text{distribution function of } X \text{ under condition } X \le z \\
F^{|z}_{|y}(x) &= [F(x)-F(y)]/[F(z)-F(y)],\ y \le x \le z,\ - \text{distribution function of } X \text{ under condition } y < X \le z \\
\mu = \mu_F &= \int_0^1 F^{-1}(x)\,dx - \text{mean of } F \\
m^2 = m_F^2 &= \int_0^1 [F^{-1}(x)]^2\,dx - \text{second raw moment of } F \\
\sigma^2 = \sigma_F^2 &= m_F^2 - \mu_F^2 - \text{variance of } F \\
\varsigma = \varsigma_F &- \int_0^1 |F^{-1}(x) - \mu_F|\,dx - \text{mean absolute deviation of } F \\
\mu_{F^0_{|y}} &- \text{mean of } F^0_{|y} \\
m^2_{F^0_{|y}} &- \text{second raw moment of } F^0_{|y} \\
M^2 = M^2_{F^0_{|y}} &= [\sigma_F^2 + (\mu_F - y)^2]/[1-F(y)] \\
\mathcal{P}_n(F) &- \text{family of all distributions on } \Re^n \text{ with common marginal } F \\
U(x) &= x,\ 0 \le x \le 1,\ - \text{standard uniform distribution function} \\
V(x) &= 1 - \exp(-x),\ x \ge 0,\ - \text{standard exponential distribution function} \\
W(x) &- \text{a (distinguished) distribution function} \\
a = a_W &- \text{(finite) left endpoint of (interval) support of } W \\
d = d_W &- \text{right endpoint of (interval) support of } W \\
w(x) &- \text{density function of } W,\ \text{weight function in } L^2([a_W, d_W], w(x)dx) \\
\mu_W(\beta) &= E_W(X|X > \beta) \\
\sigma_W^2(\beta) &= \text{Var}_W(X|X > \beta) \\
\check\mu_W(\beta) &= E_W(X|X \le \beta) \\
\check\sigma_W^2(\beta) &= \text{Var}_W(X|X \le \beta)
\end{aligned}
$$

1.2 Notation

$$\bar{\mu}_W(\beta) = E_W \min\{X, \beta\}$$
$$\bar{\nu}_W^2(\beta) = E_W[\min\{X, \beta\}]^2$$
$$\bar{\sigma}_W^2(\beta) = \mathrm{Var}_W \min\{X, \beta\}$$
$$\eta_W(\alpha, \beta) = E_W(X - \alpha)\mathbf{1}_{[\beta,d)}(X)$$
$$\vartheta_W^2(\alpha, \beta) = \mathrm{Var}_W(X - \alpha)\mathbf{1}_{[\beta,d)}(X)$$
$$\hat{\eta}_W(\beta) = \eta_W(\beta, \beta) = E_W(X - \beta)_+$$
$$\hat{\nu}_W^2(\beta) = E_W[(X - \beta)_+]^2$$
$$\hat{\vartheta}_W^2(\beta) = \vartheta_W^2(\beta, \beta) = \mathrm{Var}_W(X - \beta)_+$$
$$\check{\eta}_W(\beta) = E_W(\beta - X)_+$$
$$\check{\vartheta}_W^2(\beta) = \mathrm{Var}_W(\beta - X)_+$$

$X_1, \ldots, X_n, \ldots$ — independent identically distributed (i.i.d.) random variables with distribution function F

$Y_1, \ldots, Y_n, \ldots$ — possibly dependent identically distributed random variables with common marginal F

$X_{j:n}$ — jth order statistic of $X_1, \ldots, X_n$, $1 \leq j \leq n$

$Y_{j:n}$ — jth order statistic of $Y_1, \ldots, Y_n$, $1 \leq j \leq n$

$\sum_{j=1}^n c_j X_{j:n}$ — L-statistic of independent sample

$\sum_{j=1}^n c_j Y_{j:n}$ — L-statistic of dependent sample

$\mathbf{c} = (c_1, \ldots, c_n)$ — vector of coefficients of L-statistic

L_n — nth occurrence time of (first) record (increase in sequence of sample maxima $X_{j:j}$, $j \geq 1$)

$R_n = X_{L_n}$ — nth value of (first) record,

$L_n^{(k)}$ — nth occurrence time of (kth) record (increase in sequence of kth greatest order statistics $X_{j+1-k:j}$, $j \geq k$), $k \geq 1$

$R_n^{(k)} = X_{L_n^{(k)}+1-k:L_n^{(k)}}$ — nth value of kth record

$f_{j:n}(x) = n\binom{n-1}{j-1}x^{j-1}(1-x)^{n-j}\mathbf{1}_{[0,1]}(x)$ — density function of jth order statistic of i.i.d. standard uniform sample of size n, $1 \leq j \leq n$, expectation functional for $X_{j:n}$

$F_{j:n}(x) = \sum_{k=j}^n \binom{n}{k}x^k(1-x)^{n-k}$, $0 \leq x \leq 1$, — distribution function of $f_{j:n}(x)$

$\tilde{G}_{j:n}(x)$ — distribution function of jth order statistic of dependent identically distributed sample of size n

$G_{j:n}(x) = (nx+1-j)\mathbf{1}_{[(j-1)/n,1]}(x)/(n+1-j)$ — stochastically largest distribution function of jth order statistic of dependent sample of size n with standard uniform marginal

$g_{j:n}(x) = n\mathbf{1}_{[(j-1)/n,1)}(x)/(n+1-j)$ — density function of $G_{j:n}$, expectation functional for $Y_{j:n}$

$G_\mathbf{c}(x)$ — the greatest convex function satisfying $G(j/n) \leq \sum_{k=1}^{j} c_k$, $0 \leq j \leq n$, $0 \leq x \leq 1$

$g_\mathbf{c}(x) = \sum_{j=1}^{n} d_j \mathbf{1}_{[(j-1)/n, j/n)}(x)$ — (right) derivative of $G_\mathbf{c}(x)$, expectation functional for $\sum_{j=1}^{n} c_j Y_{j:n}$

$f_n(x) = [-\ln(1-x)]^n \mathbf{1}_{[0,1)}(x)/n!$ — density function of nth value of (first) record of i.i.d. standard uniform sequence, expectation functional of R_n

$F_n(x) = (1-x) \sum_{j=0}^{n} [-\ln(1-x)]^j/j!$, $0 \leq x \leq 1$, — distribution function of $f_n(x)$

$f_n^{(k)}(x) = k^{n+1}[-\ln(1-x)]^n (1-x)^{k-1} \mathbf{1}_{[0,1)}(x)/n!$ — density function of nth value of kth record of i.i.d. standard uniform sequence, expectation functional for $R_n^{(k)}$

$F_n^{(k)}(x) = (1-x)^k \sum_{j=0}^{n} k^j [-\ln(1-x)]^j/j!$, $0 \leq x \leq 1$, — distribution function of $f_n^{(k)}(x)$

$\preceq_c$ — convex order of distribution functions: $F \succeq_c W$ if $F^{-1}W(x)$ is convex on $[a_W, d_W)$

$\preceq_*$ — star order of distribution functions: $F \succeq_* W$ if $F^{-1}W(x)$ is starshaped; that is, $F^{-1}W(x)/(x - a_W)$ is nondecreasing on $[a_W, d_W)$

$\preceq_s$ — s-order of symmetric distribution functions: $F \succeq_s W$ if $F^{-1}W(x)$ is convex on $[\mu_W, d_W)$

$\mathcal{C}^\nearrow = \{g \in L^2([0,1), dx) : g \text{ is nondecreasing}\}$ — family of quantile functions $F^{-1}(x)$ of arbitrary distribution functions $F(x)$ with finite variance

$\mathcal{C}^0 = \{g \in \mathcal{C}^\nearrow : \int_0^1 g(x)\, dx = 0\}$ — family of respective centered quantile functions $F^{-1}(x) - \mu_F$

$\mathcal{C}^+ = \{g \in \mathcal{C}^\nearrow : g(0) = 0\}$ — family of quantile functions of life distributions with $a_F = 0$

$\mathcal{C}^s = \{g \in \mathcal{C}^\nearrow : g(x) = -g(1-x-)\}$ — family of centered quantile functions of symmetric distributions

$\mathcal{C}^\bullet$ — any of $\mathcal{C}^\nearrow, \mathcal{C}^0, \mathcal{C}^+, \mathcal{C}^s$

$\mathcal{C}_W^\bullet = \{gW : g \in \mathcal{C}^\bullet\} \subset L^2([a_W, d_W), w(x)dx)$ — family of compositions of $W(x)$ with (centered) quantile functions of $\mathcal{C}^\bullet$ (with a_W replaced by μ_W in the last case)

$\mathcal{C}_{\succeq_c(\preceq_c)W}^\bullet = \{g \in \mathcal{C}_W^\bullet : g \text{ is convex (concave) on } [a_W, d_W)\}$ — family of compositions of $W(x)$ with (centered) quantile functions of $\mathcal{C}^\bullet$ for $F \succeq_c (\preceq_c) W$

$\mathcal{C}_{\succeq_*(\preceq_*)W}^\bullet = \{g \in \mathcal{C}_W^\bullet : g \text{ is (anti)starshaped on } [a_W, d_W)\}$ — family of compositions of $W(x)$ with (centered) quantile functions of $\mathcal{C}^\bullet$ for $F \succeq_* (\preceq_*) W$

$\mathcal{C}^s_{\succeq_s(\preceq_s)W}$ = $\{g \in \mathcal{C}^s_W : g$ is convex (concave) on $[\mu_W, d_W]\}$
— family of compositions of symmetric $W(x)$ with centered quantile functions of $\mathcal{C}^s$ for symmetric $F \succeq_s (\preceq_s) W$

In particular:

$\mathcal{C}^\bullet_{\succeq_c(\preceq_c)U}$ — (centered) quantile functions of distributions with decreasing (increasing) density

$\mathcal{C}^\bullet_{\succeq_*(\preceq_*)U}$ — (centered) quantile functions of distributions with decreasing (increasing) density on the average

$\mathcal{C}^s_{\succeq_s(\preceq_s)U}$ — centered quantile functions of symmetric unimodal (U-shaped) distributions

$\mathcal{C}^\bullet_{\succeq_c(\preceq_c)V}$ — compositions of $V(x) = 1 - \exp(-x)$ with (centered) quantile functions of distributions with decreasing (increasing) failure rate

$\mathcal{C}^\bullet_{\succeq_*(\preceq_*)V}$ — compositions of $V(x) = 1 - \exp(-x)$ with (centered) quantile functions of distributions with decreasing (increasing) failure rate on the average

$\mathcal{C}^\bullet_\bullet$ — any of above defined convex cones

$P^\bullet_\bullet$ — projection onto $\mathcal{C}^\bullet_\bullet$

$A = A^\bullet_\bullet(p)$ — sharp bound on $F^{-1}(p)$ determined by projection onto $\mathcal{C}^\bullet_\bullet$

$B = B^\bullet_\bullet(j, n)$ — sharp bound on expectation of $X_{j:n}$ (independent case) determined by projection onto $\mathcal{C}^\bullet_\bullet$

$C = C^\bullet_\bullet(j, n)$ — sharp bound on expectation of $Y_{j:n}$ (dependent case) determined by projection onto $\mathcal{C}^\bullet_\bullet$

$D = D^\bullet_\bullet(k, n)$ — sharp bound on expectation of $R_n^{(k)}$ determined by projection onto $\mathcal{C}^\bullet_\bullet$

2
Basic Notions

2.1 Elements of Hilbert Space Theory

We recall here some basic facts about the Hilbert spaces that are used in the sequel. They can be found in textbooks on functional analysis (see, e.g., Balakrishnan [9]). A pair $(\mathcal{H}, (\cdot, \cdot))$ is called a real *inner product space* if $\mathcal{H}$ is a real linear space and the function $(\cdot, \cdot) : \mathcal{H} \times \mathcal{H} \mapsto \Re$, referred to further as the *inner product*, is linear in each argument when the other is fixed, symmetric under rearrangement of arguments, and positive if both arguments are identical and nonzero. These properties imply the *Schwarz inequality*

$$\forall g, h \in \mathcal{H} \quad (g, h) \leq [(g, g)(h, h)]^{1/2}. \tag{2.1}$$

This is trivial when either of the arguments is zero. Otherwise we conclude (2.1) from the relations

$$0 \leq \left(g - \frac{(g,h)}{(h,h)}h, g - \frac{(g,h)}{(h,h)}h\right)(h,h) = (g,g)(h,h) - (g,h)^2.$$

This also shows that (2.1) becomes the equality iff $g = 0$, $h = 0$, or $g = \alpha h$ for some $\alpha > 0$. We use (2.1) for verifying that the function

$$h \mapsto ||h|| = (h, h)^{1/2}$$

defines a *norm* in $\mathcal{H}$. If $(\mathcal{H}, ||\cdot||)$ is complete, then $(\mathcal{H}, (\cdot, \cdot))$ is called the *Hilbert space*. The Riesz representation theorem asserts that every linear continuous functional defined on a Hilbert space can be written as

$$T_h(g) = (g, h), \quad g \in \mathcal{H},$$

for some $h \in \mathcal{H}$. By (2.1) again, $||T_h|| = ||h||$. One can see that the normalized nonzero functional $T_h(g)/||g||$, $0 \neq g \in \mathcal{H}$, attains its maximum $||h||$ at $g = \alpha h$ with $\alpha > 0$.

In numerous statistical problems, it is important to maximize a linear normalized functional over a convex cone in a Hilbert space. We say that $\mathcal{C} \subset \mathcal{H}$ is a *convex cone* if $f, g \in \mathcal{C}$ implies that $\alpha f + \beta g \in \mathcal{C}$ for arbitrary $\alpha, \beta \geq 0$. If $h \in \mathcal{C}$, then the solution of our restricted maximization problem coincides with that of the general one. Otherwise we show that h should be replaced by its *projection* Ph onto $\mathcal{C}$, that is, the element of $\mathcal{C}$ that is least distant from h. This can be deduced from the following theorem (cf. Balakrishnan [9, Section 1.4]).

Theorem 1 *If h is an arbitrary element of a real Hilbert space $\mathcal{H}$ and $\mathcal{C}$ is a closed convex cone in $\mathcal{H}$, then there exists a unique projection Ph of h onto $\mathcal{C}$ that is characterized by two relations*

$$\forall\, g \in \mathcal{C} \quad (g, h) \leq (g, Ph), \tag{2.2}$$
$$(Ph, h) = (Ph, Ph). \tag{2.3}$$

This is a refinement of the statement that there is a uniquely defined projection $\bar{P}h$ of arbitrary $h \in \mathcal{H}$ onto a closed convex set $\bar{\mathcal{C}} \subset \mathcal{H}$, and $\bar{P}h$ satisfies

$$\forall\, g \in \bar{\mathcal{C}} \quad (g, h - \bar{P}h) \leq (\bar{P}h, h - \bar{P}h) \tag{2.4}$$

(see Balakrishnan [9, Section 1.4]). The projection is the only point of $\bar{\mathcal{C}}$ that satisfies (2.4).

Observe that for $Ph \neq \mathbf{0}$ relation (2.2) combined with (2.1) gives

$$\forall\, \mathbf{0} \neq g \in \mathcal{C} \quad T_h(g)/||g|| \leq ||Ph|| > 0. \tag{2.5}$$

Setting $g = \alpha Ph$ for some positive α and using (2.3), the equality holds in (2.5). If $Ph = \mathbf{0}$ then, due to (2.2), T_h is nonpositive on $\mathcal{C}$, and clearly $T_h(Ph) = 0$. Due to the fundamental significance of Theorem 1 for our further considerations, we recall its proof here.

PROOF OF THEOREM 1. We first show that for every $h \in \mathcal{H}$ there exists a unique $Ph \in \mathcal{C}$ that minimizes distance $||g - h||$ over all $g \in \mathcal{C}$. The claim is trivial if $h \in \mathcal{C}$. Otherwise there exists a sequence $g_n \in \mathcal{C}$, $n \geq 1$, such that

$$\lim_{n \to \infty} ||g_n - h|| = \inf_{g \in \mathcal{C}} ||g - h|| = D > 0,$$

say. For arbitrary two elements of the sequence we have

$$\left|\left|\frac{g_n - h}{2} - \frac{g_m - h}{2}\right|\right|^2 + \left|\left|\frac{g_n - h}{2} + \frac{g_m - h}{2}\right|\right|^2$$
$$= \frac{1}{2}(||g_n - h||^2 + ||g_m - h||^2). \tag{2.6}$$

For arbitrary $\epsilon > 0$, the right-hand side of (2.6) is not greater than $D^2 + \epsilon$ if n and m are large enough. Since $(g_n + g_m)/2 \in \mathcal{C}$, the latter term of the left-hand side is not less than D^2. Therefore we have

$$\|(g_n - h)/2 - (g_m - h)/2\|^2 \leq \epsilon,$$

which, by completeness of $\mathcal{H}$ and closedness of $\mathcal{C}$, implies that g_n, $n \geq 1$, has a unique limit $Ph \in \mathcal{C}$, say. Relations

$$|\|g_n - h\| - \|Ph - h\|| \leq \|g_n - Ph\| \to 0, \quad \text{as } n \to \infty,$$

the former being concluded from the triangle inequality, imply that actually $\|Ph - h\| = D$.

Now we verify that (2.4) completely characterizes Ph. For arbitrary $g \in \mathcal{C}$ we define a function D_g on the unit interval $[0,1]$ as

$$\begin{aligned} D_g(\alpha) &= \|h - (1-\alpha)Ph - \alpha g\|^2 \\ &= \|h - Ph\|^2 + 2\alpha(h - Ph, Ph - g) + \alpha^2\|Ph - g\|^2. \end{aligned}$$

We have

$$D'_g(0) = 2(h - Ph, Ph - g), \tag{2.7}$$

$$D''_g(\alpha) = 2\|Ph - g\|^2 \geq 0. \tag{2.8}$$

By (2.8), relation (2.4), identical with nonnegativity of (2.7), is the necessary and sufficient condition for the nondecreasing of both D'_g and D_g. Therefore

$$D_g(0) = \|h - Ph\|^2 \leq D_g(1) = \|h - g\|^2, \quad g \in \mathcal{C}.$$

Now we use the fact that $\mathcal{C}$ is a convex cone. Plugging $g = \alpha Ph$ into (2.4), we obtain

$$(\alpha - 1)(Ph, h - Ph) \leq 0, \quad \alpha \geq 0,$$

which yields $(Ph, h - Ph) = 0$. This proves (2.3), and in combination with (2.4) gives (2.2) ∎

Here we concentrate on functionals on Hilbert space $L^2([a,d), w(x)\,dx)$, for some $-\infty < a < d \leq +\infty$, $w: [a,d) \mapsto \Re_+$. The space consists of square integrable functions on interval $[a,d)$ with a positive weight function w, and the inner product is defined by

$$(g, h) = \int_a^d g(x)h(x)w(x)\,dx. \tag{2.9}$$

Here the Schwarz inequality takes on the form

$$\int_a^d g(x)h(x)w(x)\,dx \leq \left[\int_a^d g^2(x)w(x)\,dx \int_a^d h^2(x)w(x)\,dx\right]^{1/2}. \tag{2.10}$$

Below we present two examples of projections onto convex cones contained in $L^2([a,d], w(x)\,dx)$. The former has a simple solution. In the latter, the form of projection is more complicated and depends on the weight function w.

EXAMPLE 2. Let
$$\mathcal{C}_+ = \{h \in L^2([a,d], w(x)\,dx) : h \geq 0\}.$$

Verifying (2.2) and (2.3) we deduce that
$$P_+ h = h_+ = \max\{h, 0\}$$
is the projection of h onto $\mathcal{C}_+$ for every $h \in \mathcal{H}$. Also, one can check directly that h_+ is actually the nonnegative function closest to h for arbitrary weight w. ■

EXAMPLE 3. Consider the set $\mathcal{C}_W^{\nearrow}$ of all nondecreasing functions in the Hilbert space $L^2([a,d], w(x)\,dx)$. We assume that the constant and linear functions belong to the space. Then
$$W(y) = \int_a^y w(x)\,dx \tag{2.11}$$

is a finite, strictly increasing, and absolutely continuous function. The same holds for its well-defined inverse $W^{-1} : [0, W(d)) \mapsto \Re_+$. By finiteness of
$$(h, \mathbf{1}) = \int_a^d h(x) w(x)\,dx = \int_0^{W(d)} hW^{-1}(x)\,dx, \tag{2.12}$$

we can define an absolutely continuous function
$$H_W(y) = \int_0^y hW^{-1}(x)\,dx, \quad 0 \leq y < W(d), \tag{2.13}$$

and its greatest convex minorant $\bar{H}_W$, with a nondecreasing derivative $\bar{h}_W$, say.

We prove that $h_W^{\nearrow}$ defined as $h_W^{\nearrow} = \bar{h}_W W \in \mathcal{C}_W^{\nearrow}$ is the projection of h onto $\mathcal{C}_W^{\nearrow}$ by checking (2.2) and (2.3). For the former one, we need the following lemma (cf. Marshall and Olkin [55, Proposition A.2.(iii), p. 444]).

Lemma 1 *If $G \leq H$ are functions of bounded variation on an interval $[A, D]$, which are equal at the endpoints, then*
$$\int_A^D g(x)\,G(dx) \geq \int_A^D g(x)\,H(dx) \tag{2.14}$$

holds for every nondecreasing function g for which both the integrals exist.

PROOF (cf. Marshall and Proschan [56]). The statement is easily verified for all indicator functions $\mathbf{1}_{[y,D)}(x)$, $A < y < D$. Therefore this is true for all positive combinations of the indicator functions, and, by the Lebesgue monotone convergence theorem, for arbitrary nonnegative nondecreasing functions as well. Therefore the reversed inequality holds for the nonnegative nonincreasing functions. Finally, we represent an arbitrary nondecreasing function as the difference of two nonnegative terms, the nonincreasing and nondecreasing ones, and apply the above statements to both parts. ∎

Since $\bar{H}_W \leq H_W$ satisfy the assumptions, for arbitrary $g \in \mathcal{C}_W^{\nearrow}$ we can write

$$\begin{aligned}
(g,h) &= \int_0^{W(d)} gW^{-1}(y) hW^{-1}(y)\, dy \\
&= \int_0^{W(d)} gW^{-1}(y)\, H_W(dy) \\
&\leq \int_0^{W(d)} gW^{-1}(y)\, \bar{H}_W(dy) \\
&= \int_a^d g(x) h_W^{\nearrow}(x) w(x)\, dx \\
&= (g, h_W^{\nearrow}).
\end{aligned} \qquad (2.15)$$

In order to derive (2.3), we first thoroughly analyze relations in pairs $H_W, \bar{H}_W$, and $hW^{-1}, \bar{h}_W$, and $h, h_W^{\nearrow}$. Observe that the open set $\{\bar{H}_W < H_W\}$ is a (possibly empty) union of countably many (at most) disjoint open intervals, $\bigcup_i (W(b_i), W(c_i))$, say. The function $\bar{H}_W$ is linear on each interval, and coincides with H_W at the endpoints. Therefore

$$\bar{h}_W(x) = \frac{H_W W(c_i) - H_W W(b_i)}{W(c_i) - W(b_i)}, \qquad x \in (W(c_i), W(b_i)),$$

and we have

$$\begin{aligned}
\int_{b_i}^{c_i} h(x) h_W^{\nearrow}(x) w(x)\, dx &= \int_{W(b_i)}^{W(c_i)} hW^{-1}(y) \bar{h}_W(y)\, dy \\
&= \frac{[H_W W(c_i) - H_W W(b_i)]^2}{W(c_i) - W(b_i)} \\
&= \int_{W(b_i)}^{W(c_i)} \bar{h}_W^2(y)\, dy \\
&= \int_{b_i}^{c_i} h_W^{\nearrow 2}(x) w(x)\, dx.
\end{aligned}$$

16   2. Basic Notions

Thus the equality holds for the integrals over the whole of the open set $W^{-1}(\{\bar{H}_W < H_W\})$. For the remaining part $W^{-1}(\{\bar{H}_W = H_W\})$ the conclusion holds, since $\bar{H}_W = H_W$ implies that $h_W^{\nearrow} = h$ there. Summing up, we have

$$(h, h_W^{\nearrow}) = (h_W^{\nearrow}, h_W^{\nearrow}),$$

which is the desired conclusion. ∎

The construction of the L^2-projection onto the family of monotone functions under uniform weighting was presented in Moriguti [58]. For the general case we refer to Rychlik [85]. For simplicity, we treated elements of L^2-spaces as functions rather than equivalence classes up to almost sure equality, and we follow this convention later. For example, $h \in \mathcal{C}_+$ generally means almost sure nonnegativity, and monotonicity can be precisely defined by comparisons of integrals over the intervals of the same weight. In Examples 2 and 3 we were able to determine projections for arbitrary h, but there are no general rules for constructing projections onto other convex cones. Then we apply arguments suitable for specific functions h and cones. Usually, we first try to describe the shape of the projection in a parametric way, and then precisely determine the parameters. We finally note that more often the characterization (2.2) and (2.3) is used for determining projections of principal interest. In the problems under study we first try to find the projection by minimizing the distance of the Hilbert space point, representing a functional, to a cone. The solution of the auxiliary problem satisfies (2.2) and (2.3) in particular which are needed for estimating values of the functional over the cone.

2.2   Statistical Linear Functionals

Investigations of statistical procedures treated as functionals on distribution functions were initialized by von Mises [101]. The general theory is presented in Serfling [95] and Prakasa Rao [73]. Here we confine ourselves to some statistical functions that can be represented as linear functionals on Hilbert spaces. Assume that a random variable X has a distribution function F, finite *mean* $\mu = \mu_F$, and *second raw moment* $m^2 = m_F^2$. Changing variables, we write

$$\mu = \int_{-\infty}^{+\infty} y\, F(dy) = \int_0^1 F^{-1}(x)\, dx, \tag{2.16}$$

$$m^2 = \int_{-\infty}^{+\infty} y^2\, F(dy) = \int_0^1 [F^{-1}(x)]^2\, dx, \tag{2.17}$$

where

$$F^{-1}(x) = \sup\{y : F(y) \le x\}, \quad 0 \le x < 1, \tag{2.18}$$

is the right-continuous *quantile function* of X. We can say that m_F is the norm of F^{-1} in $L^2([0,1), dx)$, and $\mu_F = (F^{-1}, \mathbf{1})$. The family of all possible quantile functions is identical to the convex cone of (right-continuous versions of) nondecreasing functions in $L^2([0,1), dx)$. For the *variance* $\sigma^2 = \sigma_F^2$, we have

$$\sigma^2 = \int_{-\infty}^{+\infty} (y-\mu)^2 \, F(dy) = \int_0^1 [F^{-1}(x) - \mu]^2 \, dx = ||F^{-1} - \mu||^2. \quad (2.19)$$

Observe that the functions $F^{-1} - \mu_F$ form the convex cone of nondecreasing functions integrating to 0. Our purpose is to determine sharp bounds for normalized statistical functionals represented as $T(F^{-1})/m_F$ and $T(F^{-1} - \mu_F)/\sigma_F$ for general and restricted classes of quantile functions. Now we present exemplary linear functionals of statistical importance acting on quantile functions in $L^2([0,1), dx)$. Considering bounds on narrower classes of distributions, it is convenient to transform the quantile functions so that other L^2-spaces are studied.

Quantiles $F^{-1}(p)$ of order $0 < p < 1$. They characterize distributions by describing levels that divide respective populations into subsets containing desired proportions of elements. Quantiles are often used for defining critical levels of tests and interval estimates. In order to evaluate $F^{-1}(p)$ in terms of moments, we represent it as a limit of continuous linear L^2-functionals

$$F^{-1}(p) = \lim_{q \searrow p} \frac{1}{q-p} \int_p^q F^{-1}(x) \, dx = \lim_{q \searrow p} (F^{-1}, \frac{1}{q-p} \mathbf{1}_{[p,q)}), \quad (2.20)$$

where $\mathbf{1}_A$ denotes the indicator function of set A. Precisely, a distribution function may have nonunique quantiles, and (2.20) defines the *upper quantile of order p*.

Expectations of order statistics of independent samples. Let $X_1, \ldots, X_n$ be independent identically distributed random variables with a common distribution function F. Then the jth *order statistic* $X_{j:n}$, $1 \leq j \leq n$, is the jth smallest value in the sequence $X_1, \ldots, X_n$. The respective distribution function is

$$\begin{aligned} P(X_{j:n} \leq x) &= P(\text{at least } j \text{ among } X_1, \ldots, X_n \text{ are } \leq x) \\ &= \sum_{k=j}^n \binom{n}{k} F^k(x)[1-F(x)]^{n-k} \\ &= F_{j:n} F(x), \end{aligned} \quad (2.21)$$

say. Since

$$f_{j:n}(x) = F'_{j:n}(x) = n \binom{n-1}{j-1} x^{j-1}(1-x)^{n-j}, \quad (2.22)$$

we have

$$\begin{align}
E_F X_{j:n} &= \int_{-\infty}^{+\infty} x\, F_{j:n} F(dx) \\
&= \int_0^1 F^{-1}(x) f_{j:n}(x)\, dx \\
&= (F^{-1}, f_{j:n}). \tag{2.23}
\end{align}$$

Note that $F_{j:n}$ and $f_{j:n}$ are the distribution and density functions, respectively, of the jth order statistic of the standard uniform i.i.d. sample of size n. Order statistics are directly used for estimating quantiles of order j/n and describing the lifetime of the *j-out-of-n reliability system* which contains n independent identical elements and operates until at least $n+1-j$ of its elements do. We can also study the expectations of linear combinations of order statistics (so called *L-statistics*)

$$E_F \sum_{j=1}^n c_j X_{j:n} = \left(F^{-1}, \sum_{j=1}^n c_j f_{j:n}\right) \tag{2.24}$$

which have numerous applications in statistical inference. For example, those with $\sum_{j=1}^n c_j = 1$ (including the *sample mean* $1/n \sum_{j=1}^n X_{j:n}$, *sample median* $X_{(n+1)/2:n}$ for odd n, *trimmed means* $1/(n-2k) \sum_{j=k+1}^{n-k} X_{j:n}$ for $j \le n/2-1$) and ones satisfying $\sum_{j=1}^k c_j \le 0$, $1 \le k \le n-1$, and $\sum_{j=1}^n c_j = 0$ (including *sample range* $X_{n:n} - X_{1:n}$ and *sample interquartile distance* $X_{n+1-\lfloor n/4 \rfloor:n} - X_{\lfloor n/4 \rfloor:n}$) are used for estimating the location and dispersion of the population, respectively. Moreover, the projection method enables us to evaluate precisely uniform convergence rates of estimates for particular families of distributions. In the exemplary case of quantile estimation, this is possible by analyzing $E_F X_{j:n} - F^{-1}(j/n)$ and $E_F(X_{j:n} - X_{kj:kn})$ for $k > 1$. For a comprehensive treatment of the theory of order statistics we refer the reader to David [22], Arnold et al. [7], and Balakrishnan and Rao [12]. The best references for their applications are Balakrishnan and Cohen [11], and Balakrishnan and Rao [13].

Order statistics of dependent samples. Suppose that $Y_1, \ldots, Y_n$ are *possibly dependent* identically distributed random variables with a common marginal F, and $Y_{j:n}$, $1 \le j \le n$, are the respective order statistics. As in the previous case, $Y_{j:n}$ may represent the lifetime of an $(n+1-j)$-out-of-n system of elements with identical failure probability, but here each element affects the other ones somehow. It may happen in particular that $Y_1 = \ldots = Y_n$, if the damage of a single element causes the immediate damage of all remaining ones. It was shown in Rychlik [79] that for

$\mathbf{c} = (c_1, \ldots, c_n) \in \Re^n$, we have

$$\sup_{P \in \mathcal{P}_n(F)} \mathrm{E}_P \sum_{j=1}^n c_j X_{j:n} = \int_0^1 F^{-1}(x) g_{\mathbf{c}}(x) \, dx = (F^{-1}, g_{\mathbf{c}}), \qquad (2.25)$$

where $\mathcal{P}_n(F)$ denotes the family of all joint distributions P on $\Re^n$ with identical marginals F, and $g_{\mathbf{c}}$ is (the right) derivative of function $G_{\mathbf{c}}$, being the greatest convex one satisfying

$$G_{\mathbf{c}}(0) = 0, \qquad G_{\mathbf{c}}(j/n) \le \sum_{i=1}^j c_i, \qquad j = 1, \ldots, n. \qquad (2.26)$$

Formula (2.25) is valid for arbitrary coefficients $c_1, \ldots, c_n$, and distribution function F with a finite expectation. The supremum is attained for some distributions in $\mathcal{P}_n(F)$. A detailed characterization of the distributions as well as arguments leading to (2.25) are presented in Section 5.1. Maximizing (2.25) over a family of marginals F, we first determine the F providing the extreme value of $(F^{-1}, g_{\mathbf{c}})$, and then take the joint distributions in $\mathcal{P}_n(F)$, for which the expectation of the L-statistic is actually equal to $(F^{-1}, g_{\mathbf{c}})$ for the specified F. By definition, $g_{\mathbf{c}}$ is a nondecreasing step function with $n-1$ jumps at most, located at some points of the form j/n, $j = 1, \ldots, n-1$. In particular, for arbitrary $1 \le j \le n$,

$$\begin{aligned}
\sup_{P \in \mathcal{P}_n(F)} \mathrm{E}_P Y_{j:n} &= \frac{n}{n+1-j} \int_{(j-1)/n}^1 F^{-1}(x) \, dx \\
&= \left(F^{-1}, \frac{n}{n+1-j} \mathbf{1}_{[(j-1)/n, 1)} \right) \\
&= (F^{-1}, g_{j:n}), \qquad (2.27)
\end{aligned}$$

say (cf. also Caraux and Gascuel [19], Rychlik [76]). One can more easily determine projections of simple step functions presented in (2.25) and (2.27) than of polynomials appearing in respective formulae (2.24) and (2.23) for the i.i.d. case. This explains the surprising fact that we have more results and of simpler forms for arbitrarily dependent samples than for standard independent observations. Note finally that the projection method allows us to measure sensitivity of L-statistics upon dependence by evaluating

$$\sup_{P \in \mathcal{P}_n(F)} \mathrm{E}_P \sum_{j=1}^n c_j (Y_{j:n} - X_{j:n}) = \left(F^{-1}, g_{\mathbf{c}} - \sum_{j=1}^n c_j f_{j:n} \right) \qquad (2.28)$$

in chosen classes of marginal distributions. These evaluations allow statisticians to choose L-statistics that are robust against violations of independence assumptions.

Record values. Record values in numerical sequences are ones that exceed all the preceding ones. For a random sequence X_j, $j \geq 1$, *record values*, and respective *record occurrence times* L_n, $n \geq 1$, are random increasing sequences. By convention, we assume

$$L_0 = 1, \quad R_0 = X_1, \qquad (2.29)$$

and further put

$$\begin{align} L_n &= \min\{j > L_{n-1} : X_j > X_{L_{n-1}}\}, & (2.30)\\ R_n &= X_{L_n}, \quad n \geq 1. & (2.31) \end{align}$$

Due to another convention, the first value of record occurs at time 1 and equals X_1. Like extreme order statistics, records are applied in estimating strength of materials, predicting natural disasters, sport achievements, and the like. They were first studied by Chandler [20]. Comprehensive studies of records can be found in Ahsanullah [1] and Arnold et al. [8]. If observations are independent identically distributed and the distribution does not have an atom at its right support endpoint, then the sequence of records is infinite almost surely. Formulae (2.29), (2.30), and (2.31) are well defined for arbitrary original sequences, but we assume further that X_j, $j \geq 1$, are independent and have an identical continuous distribution function F, say. It is obvious that if a current value of a record is given, then the conditional distribution of the next one is identical with the distribution of the parent variable under the condition that it exceeds the actual record value. In particular, for records R_n^V of an i.i.d. sequence X_j^V, $j \geq 1$, with the standard *exponential distribution function*

$$V(x) = 1 - \exp(-x), \quad x \geq 0,$$

we have

$$\begin{align} P(R_{n+1}^V - R_n^V > y | R_n^V = x) &= P(X_1^V > y + x | X_1^V > x) \\ &= \exp(-y) \\ &= P(X_1^V > y) \qquad (2.32) \end{align}$$

for arbitrary $x, y \geq 0$. It follows that the record increments $R_0^V, R_1^V - R_0^V, \ldots, R_{n+1}^V - R_n^V, \ldots$ are independent standard exponential random variables, and R_n^V has the gamma distribution $\Gamma(n+1, 1)$ with shape parameter $n+1$ and scale 1. The transformation

$$X_j = F^{-1}V(X_j^V), \quad j \geq 1,$$

produces an i.i.d. sequence with common distribution function F. By strict monotonicity, it preserves the record occurrence times. Therefore

$$R_n = X_{L_n} = F^{-1}V(X_{L_n}^V) = F^{-1}V(R_n^V), \qquad (2.33)$$

and

$$\begin{aligned}
E_F R_n &= E_V F^{-1} V(R_n^V) \\
&= \int_0^\infty F^{-1}(1-e^{-x}) \frac{x^n}{n!} e^{-x} \, dx \\
&= \int_0^1 F^{-1}(x) f_n(x) \, dx \\
&= (F^{-1}, f_n) \qquad (2.34)
\end{aligned}$$

with

$$f_n(x) = [-\ln(1-x)]^n / n!,$$

which is a desired inner product representation.

*k*th *record values.* An increasing sequence of record values arises from a nondecreasing sequence of sample maxima $(X_{n:n})$, $n \geq 1$, by crossing out all repetitions. For arbitrary fixed k, the sequence of kth greatest order statistics $(X_{n+1-k:n})$, $n \geq k$, is nondecreasing as well. By analogy, we can define *occurrence times and values of kth records* in the following way

$$\begin{aligned}
L_0^{(k)} &= k, & R_0^{(k)} &= X_{1:k}, & (2.35) \\
L_n^{(k)} &= \min\{j > L_{n-1}^{(k)} : X_j > X_{L_{n-1}^{(k)}+1-k:L_{n-1}^{(k)}}\}, & & & (2.36) \\
R_n^{(k)} &= X_{L_n^{(k)}+1-k:L_n^{(k)}}, & n &\geq 1. & (2.37)
\end{aligned}$$

The kth records were introduced by Dziubdziela and Kopociński [26]. There is also another convention of defining kth record occurrence times that consists in subtracting $k-1$ from $L_n^{(k)}$ defined in (2.35) and (2.36). This implies that we start counting records only when the first k observations are carried out. In particular, the first value of the kth record occurs at moment 1 then. However, we are concerned with record values here, which are not affected by the particular definitions of occurrence times. In contrast with standard records, a random variable $X_{L_n^{(k)}}$ observed at the kth record time $L_n^{(k)}$ for $k \geq 2$ does not necessarily become the kth record value immediately. Generally, we have

$$X_{L_n^{(k)}} = X_{L_n^{(k)}+1-i:L_n^{(k)}} \geq X_{L_n^{(k)}+1-k:L_n^{(k)}}$$

for some $1 \leq i \leq k$. Thus it becomes the ith record first and will become the kth record when $k-i$ greater values occur in the sequence. A key relation that allows us to establish the distribution function of kth records is

$$P(R_{n+1}^{(k)} > y | R_n^{(k)} = x) = \left[\frac{1-F(y)}{1-F(x)}\right]^k, \qquad y > x, \qquad (2.38)$$

(cf. (2.32)), which means that the distribution of the next value of the kth record, when the current one is known, is the same as the distribution of the minimum of k original variables X_j under the condition that they are greater than the current record. Relation (2.38) shows that distributions of nth values of kth records from the i.i.d. sample with distribution function F and standard first records of the i.i.d. sample with distribution function

$$F_{1:k}F(x) = 1 - [1 - F(x)]^k$$

(i.e., that of $X_{1:k}$) coincide. Therefore, by (2.34),

$$\begin{aligned} E_F R_n^{(k)} &= E_V (F_{1:k}F)^{-1} V(R_n^V) \\ &= \int_0^\infty F^{-1}(1 - e^{-x/k}) \frac{x^n}{n!} e^{-x}\, dx \\ &= \int_0^1 F^{-1}(x) f_n^{(k)}(x)\, dx \\ &= (F^{-1}, f_n^{(k)}), \end{aligned} \qquad (2.39)$$

where

$$f_n^{(k)}(x) = \frac{k^{n+1}}{n!}[-\ln(1-x)]^n (1-x)^{k-1}, \qquad k \geq 1,\ n \geq 0, \qquad (2.40)$$

is the density function of the nth value of the kth record of the i.i.d. standard uniform sequence, and $f_n^{(1)} = f_n$ defined in (2.34). Formula (2.38) is true for arbitrary F. However, for the discontinuous F function $F^{-1}F_{1:k}^{-1}V$ is not strictly increasing and may transform a record in the exponential sequence into a number equalizing a previous score. Therefore (2.39) is true only for continuous F.

kth lower record values. The consecutive decreasing elements $S_n^{(k)}$, $n \geq 1$, of the nonincreasing sequence of kth order statistics $X_{k:i}$, $i \geq k$, are called kth lower record values. If X_i, $i \geq 1$, are i.i.d. standard uniform, so are $1 - X_i$, $i \geq 1$. If the upper records $R_n^{(k)}$ occur in the original sequence, we observe the lower records $S_n^{(k)}$ in the latter. If F is continuous, then transformation F^{-1} provides observations with distribution function F, and preserves their order. Therefore we have

$$S_n^{(k)} = F^{-1}(1 - R_n^{(k)}),$$

and, by (2.39) and (2.40),

$$\begin{aligned} E_F S_n^{(k)} &= \int_0^1 F^{-1}(x) f_n^{(k)}(1-x)\, dx \\ &= \int_0^1 F^{-1}(x) \frac{k^{n+1}}{n!} (-\ln x)^n x^{k-1}\, dx. \end{aligned}$$

Conditional expectations of order statistics. Suppose that the common distribution function F of $X_1, \ldots, X_n$ is absolutely continuous. Then the conditional distribution of $X_{j:n}$ given that $X_{i:n} = y$ for some $1 \leq i < j \leq n$ is the same as the distribution of the $(j-i)$th order statistic obtained from a sample of size $n-i$ from a population with distribution function

$$F_{|y}(x) = \frac{F(x) - F(y)}{1 - F(y)}, \quad x \geq y,$$

which is actually the distribution function of X_1 provided that $X_1 > y$. In fact, order statistics $X_{j:n}$, $1 \leq j \leq n$, of *continuous populations* form Markov chains with respect to j. This means that conditional distribution of $X_{j:n}$ when $X_{i_1:n} = y_1 < \ldots < X_{i_k:n} = y_k$, $1 \leq i_1 < \ldots < i_k < j \leq n$, depends only on the value of $X_{i_k:n}$. The relations are useful in predicting further failures of elements of reliability systems, and in statistical inference based on partial (truncated) observations. The proofs, based on the use of joint distributions of several order statistics, can be found in Arnold et al. [7, Section 2.4]. In particular, we have

$$E_F(X_{j:n}|X_{i_1:n} = y_1 < \ldots < X_{i_k:n} = y_k) = E_F(X_{j:n}|X_{i_k:n} = y_k)$$

for $1 \leq i_1 < \ldots < i_k < j \leq n$, and

$$\begin{aligned} &E_F(X_{j:n}|X_{i:n} = y) \\ &= E_{F_{|y}} X_{j-i:n-i} \\ &= \int_0^1 F^{-1}(F(y) + [1-F(y)]x) f_{j-i:n-i}(x)\, dx \\ &= \int_0^1 F^{-1}(x) f_{j-i:n-i}\left(\frac{x-F(y)}{1-F(y)}\right) \frac{\mathbf{1}_{[F(y),1)}(x)}{1-F(y)}\, dx \quad (2.41) \end{aligned}$$

for $1 \leq i < j \leq n$. Similar results are obtained for the reversed conditioning. The conditional distribution of the $X_{j:n}$ when $X_{k:n} = z$ is given for $k > j$ is identical with the distribution of the jth order statistic from the sample of size $k-1$ with the common distribution function of X_1 under the condition that $X_1 \leq z$ which equals

$$F^{|z}(x) = \frac{F(x)}{F(z)}, \quad x \leq z.$$

If both $X_{i:n} = y$ and $X_{k:n} = z$ for $i < j < k$ are known, then the conditional distribution of $X_{j:n}$ coincides with that of $X_{j-i:k-1-i}$ of the doubly truncated population with distribution function

$$F_{|y}^{|z}(x) = \frac{F(x) - F(y)}{F(z) - F(y)}, \quad y < x \leq z$$

(see also Arnold et al. [7, Section 2.4]). Respective analogues of (2.41) are

$$\begin{aligned}
& E_F(X_{j:n}|X_{k:n} = z) \\
=\ & E_{F|z} X_{j:k-1} \\
=\ & \int_0^1 F^{-1}(x) f_{j:k-1}\left(\frac{x}{F(z)}\right) \frac{\mathbf{1}_{[0,F(z))}(x)}{F(z)}\, dx \qquad (2.42)
\end{aligned}$$

$$\begin{aligned}
& E_F(X_{j:n}|X_{i:n} = y, X_{k:n} = z) \\
=\ & E_{F|z \atop |y} X_{j-i:k-1-i} \\
=\ & \int_0^1 F^{-1}(x) f_{j-i:k-1-i}\left(\frac{x - F(y)}{F(z) - F(y)}\right) \frac{\mathbf{1}_{[F(y),F(z))}(x)}{F(z) - F(y)}\, dx \qquad (2.43)
\end{aligned}$$

for $i < j < k$ and $y < z$. Order statistics of discontinuous populations do not have the above-mentioned properties in general. Distribution theory and dependence structure of order statistics from discrete populations were thoroughly discussed in Nagaraja [62]. Conditional expectations of functions of adjacent order statistics for general populations were examined by Franco and Ruiz [30].

Conditional expectations of record statistics. We already referred to the fact that the distribution of the $(m+1)$st record value in an atomless i.i.d. sequence X_i, $i \geq 1$, under the condition that $R_m = y$ is that of the original observation X_i under condition $X_i > y$. This is actually the record value number 0 from the sequence with distribution function $F_{|y}$. Looking for further record values, we can merely confine ourselves to the Xs that exceed level y. The $(m+2)$nd record value is the first one among Xs greater than y that is also greater than the first X_i exceeding y. In other words, this is the first record value from the population with distribution $F_{|y}$. Arguing inductively, we conclude that the distribution of R_n under the condition that $R_m = y$ for some $m < n$ is identical with the distribution of the $(n-m-1)$st record value in an i.i.d. sequence with common distribution F truncated on the left at y. Moreover, sequences of record values have the Markov structure (see, e.g., Nevzorov and Balakrishnan [64, p. 527]). Analogous properties have kth record values for general k. Justifying the claim, it suffices to recall the fact that distributions of the kth and first record values from distributions F and $F_{1:k}F = 1 - (1-F)^k$, respectively, coincide. One can easily check that

$$\begin{aligned}
(F_{1:k}F)_{|y}(x) &= [1 - (1-F)^k]_{|y}(x) \\
&= 1 - \left[\frac{1-F(x)}{1-F(y)}\right]^k \\
&= 1 - [1 - F_{|y}(x)]^k \\
&= F_{1:k}(F_{|y})(x).
\end{aligned}$$

Accordingly, for continuous parent distributions F, $1 \leq m < n$, and $k \geq 1$, we have

$$\begin{aligned}
E_F(R_n^{(k)}|R_m^{(k)} = y) &= E_{F_{1:k}F}(R_n|R_m = y) \\
&= E_{F_{1:k}F_{|y}} R_{n-m-1} \\
&= E_{F_{|y}} R_{n-m-1}^{(k)} \\
&= \int_0^1 F^{-1}(x) f_{n-m-1}^{(k)}\left(\frac{x - F(y)}{1 - F(y)}\right) \frac{\mathbf{1}_{[F(y),1)}(x)}{1 - F(y)} dx \\
&= \int_0^1 F^{-1}(x) \frac{k^{n-m}(1-x)^{k-1}}{(n-m-1)![1-F(y)]^k} \\
&\quad \times \left[\ln \frac{1-F(y)}{1-x}\right]^{n-m-1} \mathbf{1}_{[F(y),1)}(x) dx. \quad (2.44)
\end{aligned}$$

Predictions of further record values are of vital interest in constructing protection devices. Reconstruction of previous record values is also possible. Nagaraja [61] proved that in the i.i.d. samples with continuous distribution function F, the conditional distribution of R_m under the condition $R_n = z$ is identical with the distribution of the mth order statistic from the i.i.d. sample of size $n - 1 \geq m$ with the distribution function

$$\tilde{F}^{|z}(x) = \frac{V^{-1}F(\min\{x, z\})}{V^{-1}F(z)}.$$

Accordingly, elementary calculations lead us to the linear functional representation

$$\begin{aligned}
&E_F(R_m|R_n = z) \\
&= E_{\tilde{F}^{|z}} X_{m:n-1} \\
&= \int_0^1 (\tilde{F}^{|z})^{-1}(x) f_{m:n-1}(x) \, dx \\
&= \int_0^1 F^{-1}(x) f_{m:n-1}\left(\frac{-\ln(1-x)}{-\ln(1-F(z))}\right) \frac{\mathbf{1}_{[0,F(z))}(x)}{-(1-x)\ln(1-F(z))} dx.
\end{aligned}$$

Analysis of the functional is certainly a challenging task. Franco and Ruiz [29] studied conditional expectations of functions of the nearest previous record.

2.3 Restricted Families of Distributions

We formally define the convex cone of quantile functions considered in the previous section as

$$\mathcal{C}^{\nearrow} = \{g \in L^2([0,1), dx) : g \text{ — nondecreasing, right continuous}\}. \quad (2.45)$$

Right continuity assumption is inessential for problems of L^2-projections. It is introduced here in order to preserve the consistency with definitions of quantile functions, and one can simply replace an arbitrary nondecreasing function by the right continuous version. We regard elements of convex cones defined here and statistical functionals as being right continuous. Observe that projecting functions onto (2.45) we maximize normalized statistical functionals in $||F^{-1}|| = m_F$ units. In order to get more subtle evaluations in terms of mean and variance, we should consider the class of $F^{-1} - \mu_F$ functions defined as

$$C^0 = \{g \in C^\nearrow : \int_0^1 g(x)\,dx = 0\}. \quad (2.46)$$

It is also of interest to study *life* distributions, which, by definition, have the left support endpoint at zero, and generate the following family of nonnegative quantile functions

$$C^+ = \{g \in C^\nearrow : g(0) = 0\}. \quad (2.47)$$

In this case we have another pair of location and scale parameters: the left endpoint of support $a_F = 0$ and the square root of the second moment m_F, respectively.

The family of *symmetric* distributions (about the respective expectation $\mu = \mu_F$) is characterized by equivalent relations

$$F(x - \mu) = 1 - F(\mu - x-), \quad (2.48)$$
$$F^{-1}(x) - \mu = -[F^{-1}(1 - x-) - \mu]. \quad (2.49)$$

It is convenient to study the upper halves of $F^{-1} - \mu_F$, which form the cone

$$C^s = \{g \in L^2([\tfrac{1}{2},1), 2dx) : g - \text{nondecreasing}, g(\tfrac{1}{2}) = 0\}, \quad (2.50)$$

and extend the functions to the whole unit interval using (2.49). This implies the modification of functionals which consists in symmetrizing them about $1/2$. Indeed, by (2.49), this yields

$$T_h(F^{-1} - \mu) = \int_0^1 [F^{-1}(x) - \mu] h(x)\,dx$$
$$= \int_{1/2}^1 2[F^{-1}(x) - \mu] h^s(x)\,dx \quad (2.51)$$

for

$$h^s(x) = [h(x) - h(1 - x-)]/2. \quad (2.52)$$

Observe that the norm of $F^{-1} - \mu_F$ in $L^2([1/2, 1), 2dx)$ is σ_F.

2.3 Restricted Families of Distributions

In various models, existence and monotonicity of the density function of the parent distribution F is assumed. If the density is nonincreasing (nondecreasing) then F is increasing concave (convex) on its support, and F^{-1} is increasing convex (concave, respectively). We define

$$\mathcal{C}^{\nearrow}_{\succeq_c U} = \{g \in \mathcal{C}^{\nearrow} : g - \text{convex}\}, \tag{2.53}$$

$$\mathcal{C}^{\nearrow}_{\preceq_c U} = \{g \in \mathcal{C}^{\nearrow} : g - \text{concave}\}, \tag{2.54}$$

with $\mathcal{C}^0_{\succeq_c U}$ ($\mathcal{C}^0_{\preceq_c U}$) and $\mathcal{C}^+_{\succeq_c U}$ ($\mathcal{C}^+_{\preceq_c U}$) denoting intersections of (2.53) ((2.54), respectively) with linear subspaces of functions integrating to 0 (vanishing at 0, respectively). The apparently awkward notation is justified below. We say that $F(x)$ succeeds the *standard uniform distribution function*

$$U(x) = x, \quad 0 \leq x \leq 1,$$

in the *convex order* (written as $F \succeq_c U$) if $F^{-1}U = F^{-1}$ is convex on $[0, 1)$. The reversed relation $F \preceq_c U$ is defined by the convexity of $U^{-1}F = F$ on the support of F, being equivalent with the concavity of F^{-1}. By convention, we call distributions satisfying $F \succeq_c U$ and $F \preceq_c U$ the *decreasing* and *increasing density distributions*, respectively. The *convex order*, defined in van Zwet [100], is a partial order of (absolutely) continuous distribution functions, invariant under location and scale transformations (see Dharmadhikari and Joag-dev [25, Theorem 9.1, p. 217]). Therefore we can generalize (2.53) and (2.54) as

$$\mathcal{C}^{\nearrow}_{\succeq_c W} = \{g \in L^2([a_W, d_W), w(x)dx) : g - \text{nondecreasing and convex}\}, \tag{2.55}$$

$$\mathcal{C}^{\nearrow}_{\preceq_c W} = \{g \in L^2([a_W, d_W), w(x)dx) : g - \text{nondecreasing and concave}\}, \tag{2.56}$$

by taking compositions $F^{-1}W$ for an arbitrarily fixed distribution function W with support $[a, d) = [a_W, d_W)$, and density w. We also introduce convex cones $\mathcal{C}^0_{\succeq_c W}$ ($\mathcal{C}^0_{\preceq_c W}$) and $\mathcal{C}^+_{\succeq_c W}$ ($\mathcal{C}^+_{\preceq_c W}$) by adding conditions

$$\int_a^d g(x)w(x)\,dx = 0,$$

$$g(a) = 0,$$

respectively, to definition (2.55) ((2.56), respectively). These definitions are justified by properties of compositions of F^{-1} with W. In particular, we have

$$\int_a^d [F^{-1}W(y)]^2 w(y)\,dy = \int_0^1 [F^{-1}(x)]^2\,dx < \infty, \tag{2.57}$$

$$\int_a^d [F^{-1}W(y) - \mu]w(y)\,dy = \int_0^1 [F^{-1}(x) - \mu]\,dx = 0, \tag{2.58}$$

$$F^{-1}W(a) = F^{-1}(0). \tag{2.59}$$

The families of life distributions being in convex order with exponential distribution are studied in reliability theory. Observe that $F \prec_c V$ implies convexity of the *hazard function*

$$V^{-1}F(x) = -\ln[1 - F(x)],$$

and nondecrease of

$$\lambda_F(x) = (V^{-1}F)'(x) = \frac{f(x)}{1 - F(x)}.$$

Function $\lambda_F(x)$, called the *failure (hazard) rate*, describes the infinitesimal probability of failure at a short period of time following x under the condition of surviving until x. Condition $F \prec_c V$ defines the family of *distributions with increasing failure rate* (IFR, for short). Likewise, $F \succ_c V$ coincides with nonincrease of λ_F, and defines the family of *decreasing failure rate* (DFR) *distributions*. We specify here bounds on statistical functionals restricted to IFR and DFR distributions. Since $U \prec_c V$, every DFR distribution has a decreasing density, and the increase of density implies that of the failure rate.

van Zwet [100] (see also Lawrence [49]) defined a counterpart $\prec_s$ of convex order for symmetric distributions, and called it *s-order*. We have $F \prec_s W$ for F and W symmetric about μ_F and μ_W, respectively, if $F^{-1}W(x)$ is concave for $x \geq \mu_W$. This is equivalent with convexity of $F^{-1}W$ on the lower half of the support of W, and concavity-convexity of $W^{-1}F$ on the support of F. Combining constructions leading to (2.50), (2.55), and (2.56), we define the convex cones

$$\mathcal{C}^s_{\succeq_c W} = \{g \in L^2([\mu_W, d_W), 2w(x)\,dx) : g(\mu_W) = 0,\ g - \text{nondecreasing},$$
$$\text{convex}\}, \qquad (2.60)$$

and $\mathcal{C}^s_{\preceq_c W}$ replacing the convexity assumption by concavity in (2.60). It is worth pointing out here that $F \succeq_s (\preceq_s) U$ define the classes of *symmetric unimodal (U-shaped) distributions*. Multiplying the weight function by 2 (cf. also (2.50)) allows us to confine ourselves to L^2-spaces defined by means of probabilistic measures. Obviously, the constant multiplicator does not affect projections.

The *star order* $\prec_*$ of continuous life distributions, introduced in Barlow and Proschan [15], is more general partial order than the convex one (see, e.g., Dharmadhikari and Joag-dev [25, Theorem 9.1, p. 217]). By definition, $F \prec_* W$ iff $W^{-1}F$ is *starshaped*; that is, $W^{-1}F(x)/x$ is nondecreasing on the support of F. If $F \prec_* V$ in particular, then

$$\Lambda_F(x) = \frac{-\ln[1 - F(x)]}{x} = \frac{1}{x}\int_0^x \lambda_F(y)\,dy \qquad (2.61)$$

is nondecreasing, and we can say that F has an *increasing failure rate on the average* (IFRA, for brevity). DFRA life distributions F are defined by relation $F \succeq_* V$. Also, $F \preceq_* (\succeq_*)U$ means that

$$\frac{F(x)}{x} = \frac{1}{x} \int_0^x f(y)\, dy$$

is nondecreasing (nonincreasing, respectively). Accordingly, the relation defines the family of life *distributions with increasing (decreasing) density on the average*: although f may be multimodal, the larger values in $[a_F, d_F)$ are more (less) probable than the smaller ones. The star order is scale invariant. In order to make it invariant with respect to translations as well, we generalize the definition as follows: $F \succeq_* (\preceq_*)W$ iff a_F, a_W are finite and $[F^{-1}W(x) - F^{-1}W(a_W)]/(x - a_W)$ is nondecreasing (nonincreasing) on $[a_W, d_W)$. The definition enables to establish mean-variance bounds on statistical functionals by projecting them onto convex cones of $F^{-1}W - \mu_F$ described by the generalized star relations:

$$\mathcal{C}^0_{\succeq_* W} = \{g \in L^2([a_W, d_W), w(x)dx) : g(x), \frac{g(x) - g(a_W)}{x - a_W} \text{ are nondecreasing and}$$
$$\int_{a_W}^{d_W} g(x)w(x)\, dx = 0\}, \quad (2.62)$$

$$\mathcal{C}^0_{\preceq_* W} = \{g \in L^2([a_W, d_W), w(x)dx) : g(x) \text{ is nondecreasing,}$$
$$\frac{g(x) - g(a_W)}{x - a_W} \text{ is nonincreasing,}$$
$$\int_{a_W}^{d_W} g(x)w(x)\, dx = 0\}. \quad (2.63)$$

Special emphasis is laid on cases $W = U, V$. For a detailed treatment of stochastic orders and their applications, we refer the reader to monographs of Dharmadhikari and Joag-dev [25], and Shaked and Shantikumar [96].

We adopt the convention of denoting the projection onto a given convex cone by writing P with the same upper and lower indices that appear in the notation of the cone. For example, $P^0_{\succeq_* W}h$ denotes the projection of h onto $\mathcal{C}^0_{\succeq_* W}$. Although the original quantile functions of restricted families of distributions determined by the orders discussed above form convex cones, we prefer considering compositions $F^{-1}W$. The reason is that the latter have natural analytic and geometric properties. Our process of determining the projection of a given functional consists of choosing an arbitrary starting point in the cone and constructing consecutive approximations improving the previous ones. We first try to describe the shape of the projection function and then calculate optimal parameters. It is therefore essential for the first step that we are able to check immediately if a function proposed for an approximation actually belongs to the cone. The only assumption appearing in definitions of our cones that cannot be verified at first glance is weighted integrability to 0 (see, e.g., (2.46), (2.60), (2.62), and (2.63)). One can overcome the problem using the following lemma (cf. Rychlik [87, Lemma 1]).

Lemma 2 *Suppose that C is a subset of a real Hilbert space $\mathcal{H}$ such that $g \in C$ implies $g + cg_0 \in C$ for some $g_0 \in \mathcal{H}$ and all real c. If $h_0 \in \mathcal{H}$ has a projection Ph_0 onto C, then*

$$(Ph_0, g_0) = (h_0, g_0).$$

PROOF. For arbitrary fixed $g \in C$ and $c \in \Re$, $\|g + cg_0 - h_0\|^2$ is minimized by

$$c_0 = c_0(g) = \frac{(h_0 - g, g_0)}{(g_0, g_0)}.$$

The projection Ph_0 necessarily has the form $g + c_0 g_0$ for some $g \in C$, and satisfies

$$(g + c_0 g_0, g_0) = (h_0, g_0). \blacksquare$$

Let $C_\bullet^0$ be any of the above-considered convex cones such that

$$\int_u^d g(x) w(x)\, dx = (g, \mathbf{1}) = 0$$

for all $g \in C_\bullet^0$ is assumed. Let $C_\bullet^\nearrow$ denote the extension of $C_\bullet^0$ by dropping the integral condition. Note that each $C_\bullet^\nearrow$ is translation invariant (i.e., fulfills the assumption of Lemma 2 with $g_0 = \mathbf{1}$). Therefore $(h, \mathbf{1}) = 0$ implies that $P_\bullet^\nearrow h \in C_\bullet^0$ and coincides with $P_\bullet^0 h$. Otherwise we replace h by

$$h_0 = h - \frac{(h, \mathbf{1})}{(\mathbf{1}, \mathbf{1})} \mathbf{1} = h - (h, \mathbf{1}), \qquad (2.64)$$

noting that we confined ourselves to weights generated by probability measures. Then, by (2.2) and (2.3),

$$\forall\, g \in C_\bullet^0 \quad T_h(g) = (g, h) = (g, h - (h, \mathbf{1})) = T_{h_0}(g) \leq \|P_\bullet^0 h_0\|\, \|g\|,$$
$$T_h(P_\bullet^0 h_0) = T_{h_0}(P_\bullet^0 h_0) = \|P_\bullet^0 h_0\|^2.$$

Since $(h_0, \mathbf{1}) = 0$, we have $P_\bullet^0 h_0 = P_\bullet^\nearrow h_0$. It follows that it suffices to replace functional h by (2.64) and project it onto $C_\bullet^\nearrow$, without bothering about the integral condition. In fact, translation invariance of $C_\bullet^\nearrow$ yields

$$P_\bullet^\nearrow(h_0) = P_\bullet^\nearrow(h) - (h, \mathbf{1})$$

so that we are reduced to projecting the original h onto $C_\bullet^\nearrow$, and subtracting the constant from the result.

Another problem that one should be aware of is the existence of projections and probability distributions attaining bounds on functionals in restricted families. Functional representations of expectations of records, and conditional expectations of order and record statistics are valid for absolutely continuous distributions. Moreover, the families of absolutely continuous distributions determined by the convex and star orders form convex cones that are not closed. In fact, the cones of absolutely continuous distributions are dense in the respective convex cones defined in this section. We project the functionals onto the closed convex cones, and the projections are border points of them. In many cases they correspond to distributions that are not absolutely continuous, and so do not belong to the families under study. Nevertheless, we write down formulae describing these distributions, because they enable the reader to guess easily the forms of absolutely continuous elements of restricted families with values of functionals arbitrarily close to optimal bounds. Strictly speaking, the weak convergence in the topological sense is sufficient here. However, we can easily find sequences F_k^{-1}, $k \geq 1$, of quantile functions of absolutely continuous distributions with assumed properties that tend to the desired limit F^{-1}, say, in the norm of $L^2([0,1), dx)$. Note that, by change of variables,

$$F_k^{-1} W \to F^{-1} W \quad \text{in } L^2([a_W, d_W), w(x)dx)$$

is equivalent to

$$F_k^{-1} \to F^{-1} \quad \text{in } L^2([0,1), dx).$$

More intuitively, we can express the relation in terms of random variables. If F_k, $k \geq 1$, are distribution functions of a given family, and F generated by the projection method does not belong to the family, then F_k, $k \geq 1$, attains in the limit the bound attained by F if

$$E_U[F_k^{-1}(X) - F^{-1}(X)]^2 \to 0;$$

that is, $F_k^{-1}(X)$ converges to $F^{-1}(X)$ in the mean square for a standard uniform random variable X.

For instance, (2.53) contains the quantile functions of distributions with decreasing density, and, possibly, an atom at the left endpoint a_F of support. Suppose that the solution of the projection problem, dual to a problem of evaluating a linear statistical functional over decreasing density distributions, has an atom. This can be replaced by absolutely continuous (uniform, say) components with increasing concentration about a_F so that we obtain decreasing density distributions with quantile functions tending to that of F in $L^2([0,1), dx)$. Moreover, these absolutely continuous approximations can be modified so as to preserve desired moments of limiting F.

Lemma 3 *If $H : [a, d] \mapsto \Re$ is right continuous and of bounded variation, H^- denotes its left continuous version, and $\bar{h}$ is the right derivative of the greatest convex minorant $\bar{H}$ of H, then*

$$\int_a^d g(x) \, H(dx) \leq \int_a^d g(x) \bar{h}(x) \, dx \leq \left[\int_a^d g^2(x) \, dx \int_a^d \bar{h}^2(x) \, dx \right]^{1/2} \quad (3.1)$$

for every nondecreasing function g.

The former relation in (3.1) becomes an equality iff g is constant in every interval contained in $\{\bar{H} < \min\{H, H^-\}\}$, and right (left) continuous at every discontinuity point of H (if any) such that $H > H^-$ ($H < H^-$, respectively) there. The latter relation in (3.1) becomes an equality iff either $\bar{h} = 0$ or $g = \alpha \bar{h}$ for $\alpha \geq 0$.

Also, under assumption $H(a) = H^-(d)$, function g can be replaced by arbitrary translation $g + c$ in the middle and last expressions of (3.1), and the conditions for equality.

We tacitly assume that all integrals in (3.1) are well defined and finite. Observe that the conditions for the latter equality imply those for the former, because $\bar{H}$ is linear and $\bar{h}$ is constant on every interval of the set $\{\bar{H} < \min\{H, H^-\}\}$. The arguments in the proof of Lemma 3 are similar to those used in Example 3 of Section 2.2. Since we have a uniform weighting here, introducing compositions with the antiderivative is redundant. Moreover, the first step of constructing the derivative $\bar{h}$ of the greatest convex minorant $\bar{H}$ of the antiderivative H of the projected function h also may be omitted, because in Lemma 3 we start from a distribution function H. On the other hand, this is not necessarily continuous here which results in a more sophisticated condition for equality in (3.1).

Theorem 2 (general distributions) *For all $0 < p < 1$,*

$$\frac{F^{-1}(p) - \mu_F}{\sigma_F} \leq \left(\frac{p}{1-p} \right)^{1/2}. \quad (3.2)$$

The equality in (3.2) holds iff F is the two-point distribution valued at $\mu - [(1-p)/p]^{1/2} \sigma$ and $\mu + [p/(1-p)]^{1/2} \sigma$ with probabilities p and $1-p$, respectively.

The theorem immediately follows from Lemma 3 by putting

$$H(x) = \mathbf{1}_{[p,1)}(x) - U(x)$$

with

$$\bar{h}(x) = \begin{cases} -1, & \text{if } 0 \leq x < p, \\ \frac{p}{1-p}, & \text{if } p \leq x < 1, \end{cases}$$

$$\|\bar{h}\|^2 = \frac{p}{1-p},$$

3.1 General and Symmetric Distributions

and
$$\frac{F^{-1}(x) - \mu}{\sigma} = \frac{\bar{h}(x)}{||\bar{h}||}.$$

Moriguti [58] also showed that

$$\frac{F^{-1}(q) - F^{-1}(p)}{\sigma_F} \leq \left(\frac{1}{1-q} + \frac{1}{p}\right)^{1/2}, \quad 0 < p < q < 1. \tag{3.3}$$

Special cases of (3.2) and (3.3) for the median and interquartile distance yield

$$F^{-1}\left(\frac{1}{2}\right) - \mu_F \leq \sigma_F,$$

$$F^{-1}\left(\frac{3}{4}\right) - F^{-1}\left(\frac{1}{4}\right) \leq 2^{3/2} \sigma_F.$$

Likewise for symmetric distributions and $1/2 \leq p < 1$, we can write

$$F^{-1}(p) - \mu \leq \int_{1/2}^{1} [F^{-1}(x) - \mu] \mathbf{1}_{[p,1)}(dx)$$

$$\leq \int_{1/2}^{1} [F^{-1}(x) - \mu] \overline{\mathbf{1}_{[p,1)}}'(x)\, dx$$

$$\leq 2^{-1/2} ||\overline{\mathbf{1}_{[p,1)}}'|| \sigma$$

$$= [2(1-p)]^{-1/2} \sigma, \tag{3.4}$$

because

$$\overline{\mathbf{1}_{[p,1)}}(x) = \frac{x-p}{1-p} \mathbf{1}_{[p,1)}(x)$$

is the greatest convex minorant of $\mathbf{1}_{[p,1)}$. For $p < 1/2$ we trivially obtain

$$F^{-1}(p) - \mu_F \leq F^{-1}(p) - F^{-1}\left(\frac{1}{2}\right) \leq 0.$$

Theorem 3 (symmetric distributions) *We have $F^{-1}(p) \leq \mu_F$ for every quantile of order $p < 1/2$.*

For $p \geq 1/2$, we have

$$\frac{F^{-1}(p) - \mu_F}{\sigma_F} \leq \frac{1}{[2(1-p)]^{1/2}}, \tag{3.5}$$

with the equality attained in (3.5) for the three-point distribution

$$P\left(X = \mu \pm \frac{\sigma}{[2(1-p)]^{1/2}}\right) = 1 - p,$$

$$P(X = \mu) = 2p - 1.$$

Observe that for $p = 1/2$ bounds (3.2) and (3.5) coincide, and (3.5) is sharper for $p > 1/2$.

3.2 Distributions with Monotone Density and Failure Rate

We start with the problem of evaluating the upper quantiles $F^{-1}(p)$ for all $F \succeq_c W$. Our auxiliary projection problem is to find a function in $\mathcal{C}^{\nearrow}_{\succeq_c W}$ minimizing the distance to the indicator function of the interval $[W^{-1}(p), W^{-1}(q))$ for some $q > p$. The construction is based on the following two lemmas.

Lemma 4 Let

$$h = M\mathbf{1}_{[b,c)} \in L^2([a_W, d_W), w(x)dx)$$

for $a < b < c \leq d$ and $M > 0$. Then for every $g \in \mathcal{C}^{\nearrow}_{\succeq_c W}$ there exists $g_{\alpha\beta} \in \mathcal{C}^{\nearrow}_{\succeq_c W}$ defined as

$$g_{\alpha\beta}(x) = \alpha(x - \beta)_+ \tag{3.6}$$

for some $\alpha \geq 0$, $\beta \leq b$, $\alpha(b - \beta) < M$, such that

$$\|g_{\alpha\beta} - h\| \leq \|g - h\|.$$

The two-parametric class (3.6) consists of nonnegative constants, increasing linear functions, and broken lines with two pieces: horizontal and increasing ones.

PROOF OF LEMMA 4. Note first that if $g(a) < 0$, then $g_+ = \max\{g, 0\} \in \mathcal{C}^{\nearrow}_{\succeq_c W}$ lies closer to h than the original g. If $g(a) \geq 0$ and $g(b) \geq h(b)$, then $h \leq g \neq h$ and $(g, \mathbf{1}) > (h, \mathbf{1})$. Since $\|g + C - h\|$ is minimized in the class of translations $g + C$, of a given function g by the real C satisfying $(g + C, \mathbf{1}) = (h, \mathbf{1})$ (cf. Lemma 2), we improve approximation of h by adding a negative constant and truncating at level 0.

These modifications lead us to functions satisfying $0 \leq g(a) \leq g(b) < h(b)$. Function $g = \mathbf{0}$ can be replaced by a positive constant. A correction of a nonnegative nonzero function $g \in \mathcal{C}^{\nearrow}_{\succeq_c W}$ that vanishes at c consists in setting 0 in $[c, d)$, and adding a positive constant. Therefore we can further assume that $g(c) > 0$, and define

$$\gamma = \inf\{x \in (b, c] : g(x) \geq h(x)\}.$$

We easily check that the function

$$g_\gamma(x) = \max\left\{\frac{g(\gamma) - g(b)}{\gamma - b}(x - b) + g(b), 0\right\}$$

lies between h and g, and can be alternatively parameterized as (3.6) with respective restrictions on parameters. ∎

3.2 Distributions with Monotone Density and Failure Rate

Lemma 5 *Let the assumptions of Lemma 4 hold and*

$$\int_a^d w(x)\,dx = \int_a^d h(x)w(x)\,dx = 1. \tag{3.7}$$

If

$$\frac{\int_b^c (x-a)w(x)\,dx}{\int_b^c w(x)\,dx} \le \int_a^d (x-a)w(x)\,dx, \tag{3.8}$$

then $P^{\nearrow}_{\preceq_c W} h = 1$.

Otherwise there exists a unique $\beta_* < b$ that solves

$$\frac{\int_b^c (x-\beta)w(x)\,dx}{\int_b^c w(x)\,dx} = \frac{\int_{\max\{\beta,a\}}^d (x-\beta)^2 w(x)\,dx}{\int_{\max\{\beta,a\}}^d (x-\beta)w(x)\,dx} \tag{3.9}$$

and

$$\alpha_* = \alpha_*(\beta_*) = \left[\int_{\bar\beta_*}^d (x-\beta_*)w(x)\,dx\right]^{-1} > 0 \tag{3.10}$$

such that

$$P^{\nearrow}_{\preceq_c W} h(x) = \alpha_*(x-\beta_*)_+ = \frac{(x-\beta_*)_+}{\int_{\max\{\beta_*,a\}}^d (y-\beta_*)w(y)\,dy}. \tag{3.11}$$

Precisely, if relation "$\ge$" holds in (3.9) with β replaced by a, then $\beta_* \le a$, and (3.11) is linear on $[a,d]$. In the opposite case, $\beta_* > a$ and the projection is actually a two-piece broken line.

PROOF. Our purpose is to minimize

$$\begin{aligned}
D(\alpha,\beta) &= \|g_{\alpha\beta} - h\|^2 \\
&= \alpha^2 \int_{\max\{\beta,a\}}^d (x-\beta)^2 w(x)\,dx \\
&\quad - \frac{2\alpha}{\int_b^c w(x)\,dx}\int_b^c (x-\beta)w(x)\,dx + \frac{1}{\int_b^c w(x)\,dx}
\end{aligned} \tag{3.12}$$

with respect to α and β. For fixed $\beta < b$, we easily find optimal

$$\alpha_* = \alpha_*(\beta) = \frac{\int_b^c (x-\beta)w(x)\,dx}{\int_b^c w(x)\,dx \int_{\max\{\beta,a\}}^d (x-\beta)^2 w(x)\,dx} > 0, \tag{3.13}$$

which plugged into (3.12) gives

$$D(\alpha_*(\beta),\beta) = \frac{1}{\int_b^c w(x)\,dx} - \frac{[\int_b^c (x-\beta)w(x)\,dx]^2}{[\int_b^c w(x)\,dx]^2 \int_{\max\{\beta,a\}}^d (x-\beta)^2 w(x)\,dx}. \tag{3.14}$$

38   3. Quantiles

By differentiation,

$$\begin{aligned}\frac{dD(\alpha_*(\beta),\beta)}{d\beta} &= \frac{2\int_b^c (x-\beta)w(x)\,dx}{[\int_b^c w(x)\,dx]^2 \int_{\max\{\beta,a\}}^d (x-\beta)^2 w(x)\,dx} \\ &\quad \times \left[\int_b^c w(x)\,dx \int_{\max\{\beta,a\}}^d (x-\beta)^2 w(x)\,dx \right. \\ &\quad\left. - \int_{\max\{\beta,a\}}^d (x-\beta)w(x)\,dx \int_b^c (x-\beta)w(x)\,dx\right].\end{aligned}$$

Since the factor in the first line is positive, the sign of the derivative is identical with that of the last two which is denoted by $K(\beta)$. Observe that

$$\begin{aligned}\frac{K(b)}{\int_b^d w(x)\,dx \int_b^c w(x)\,dx} &= E_W((X-b)^2|X>b)\\ &\quad - E_W(X-b|X>b)E_W(X-b|b<X<c)\\ &\geq \operatorname{Var}_W(X|X>b) > 0 \qquad (3.15)\end{aligned}$$

for a random variable X with distribution function W. Furthermore,

$$\begin{aligned}K'(\beta) &= \int_{\max\{\beta,a\}}^d w(x)\,dx \int_b^c (x-\beta)w(x)\,dx \\ &\quad - \int_b^c w(x)\,dx \int_{\max\{\beta,a\}}^d (x-\beta)w(x)\,dx\end{aligned}$$

evaluated at $\beta = b$ satisfies

$$\frac{K'(b)}{\int_b^d w(x)\,dx \int_b^c w(x)\,dx} = E_W(X|b<X<c) - E_W(X|X>b) < 0. \quad (3.16)$$

Finally,

$$K''(\beta) = -w(\beta)\int_b^c (x-\beta)w(x)\,dx \leq 0 \qquad (3.17)$$

under the convention that $w(\beta) = 0$ for $\beta < a_W$. By (3.17) and (3.16), $K'(\beta)$ is constant for $\beta < a$ and nonincreasing to $K'(b) < 0$ for $a \leq \beta < b$.

If $K'(a) \leq 0$, which means that (3.8) holds, then $K'(\beta) \leq 0$ and $K(\beta)$ is nonincreasing and, by (3.15), positive for all $\beta \leq b$. Therefore (3.14) is minimized at $\beta = -\infty$, which implies that the projection is a constant. This amounts to $(h, \mathbf{1}) = 1$, which proves the first assertion.

If $K'(a) > 0$, then $K(\beta)$ changes its sign once from minus to plus at some $\beta < b$, at which (3.14) is minimized. Equations (3.9) and $K(\beta) = 0$ are equivalent, and allow us to rewrite (3.13) as (3.10). Note that $K(a) \geq 0$

3.2 Distributions with Monotone Density and Failure Rate

implies that the solution to (3.9) satisfies $\beta \leq a$, and that the projection is linear. In the opposite case a broken line with break at $\beta > a$ is obtained. This proves the final claim of Lemma 5. ∎

Using (3.11) and writing $\underline{\beta}_* = \max\{\beta_*, a\}$, we obtain

$$\|P^0_{\succeq_c W} h - \mathbf{1}\|^2 = \|P^\nearrow_{\succeq_c W} h\|^2 - 1$$

$$= \frac{\int_{\underline{\beta}_*}^d (x - \beta_*)^2 w(x)\, dx - [\int_{\underline{\beta}_*}^d (x - \beta_*) w(x)\, dx]^2}{[\int_{\underline{\beta}_*}^d (x - \beta_*) w(x)\, dx]^2}, \qquad (3.18)$$

$$\frac{P^0_{\succeq_c W} h(x) - 1}{\|P^0_{\succeq_c W} h - \mathbf{1}\|}$$

$$= \frac{(x - \beta_*)_+ - \int_{\underline{\beta}_*}^d (y - \beta_*) w(y)\, dy}{\left\{ \int_{\underline{\beta}_*}^d (y - \beta_*)^2 w(y)\, dy - [\int_{\underline{\beta}_*}^d (y - \beta_*) w(y)\, dy]^2 \right\}^{1/2}}. \qquad (3.19)$$

Letting $c \searrow b$, we reduce the right-hand sides of (3.8) and (3.9) to $b - a$ and $b - \beta$, respectively. Analyzing the resulting formulae, we derive bounds for quantiles. We introduce some notation before presenting them. For a random variable X with distribution function W, and arbitrary $\beta \in (a, d)$, denote the expectation and variance of X under the condition of exceeding level β by

$$\mu_W(\beta) = E_W(X|X > \beta) = \frac{\int_\beta^d x w(x)\, dx}{\int_\beta^d w(x)\, dx}, \qquad (3.20)$$

$$\sigma_W^2(\beta) = \operatorname{Var}_W(X|X > \beta)$$

$$= \frac{\int_\beta^d x^2 w(x)\, dx - \mu_W^2(\beta)}{\int_\beta^d w(x)\, dx}. \qquad (3.21)$$

We also define a distance between the pth quantile and the conditional mean in respective standard deviation units

$$\delta_W(\beta) = \frac{W^{-1}(p) - \mu_W(\beta)}{\sigma_W(\beta)}. \qquad (3.22)$$

Theorem 4 $(F \succeq_c W)$ If $W^{-1}(p) \leq \mu_W$, then $F^{-1}(p) \leq \mu_F$. If $\mu_W < W^{-1}(p) \leq \mu_W + \sigma_W^2/(\mu_W - a_W)$, then

$$\frac{F^{-1}(p) - \mu_F}{\sigma_F} \leq \delta_W(a_W) = \frac{W^{-1}(p) - \mu_W}{\sigma_W}, \qquad (3.23)$$

and the equality holds for the location-scale transformations of W, that is, for

$$F(x) = W\left(\mu_W + \sigma_W \frac{x - \mu}{\sigma}\right). \qquad (3.24)$$

Finally, if $W^{-1}(p) > \mu_W + \sigma_W^2/(\mu_W - a_W)$, then there exists a unique solution $\beta_* = \beta_*(p) \in (a_W, W^{-1}(p))$ to equation

$$[W^{-1}(p) - \mu_W(\beta)][\mu_W(\beta) - \beta] = \sigma_W^2(\beta), \qquad (3.25)$$

and

$$\frac{F^{-1}(p) - \mu_F}{\sigma_F} \leq A = A_{\succeq_c W}^0(p) = \left[\frac{\delta_W^2(\beta_*) + W(\beta_*)}{1 - W(\beta_*)}\right]^{1/2}. \qquad (3.26)$$

The bound in (3.26) is attained by

$$F(x) = W\left(\beta_* + [\mu_W(\beta_*) - \beta_*][1 - W(\beta_*)]\left(1 + A\tfrac{x-\mu}{\sigma}\right)\right) \mathbf{1}_{[\mu - \sigma/A, \infty)}(x). \qquad (3.27)$$

Distribution function (3.27) has a jump of height $W(\beta_*) < p$ at the left endpoint, and shares the shape of W on its support. This is a life distribution if

$$\mu_F A_{\succeq_c W}^0(p) = \sigma_F.$$

Mean-variance bounds for $F \succeq_c U, V$ are specified in Propositions 1 and 2, respectively.

Proposition 1 (decreasing density) *If $0 < p \leq 1/2$, then $F^{-1}(p) \leq \mu_F$ holds.*

If $1/2 < p \leq 2/3$, then

$$\frac{F^{-1}(p) - \mu_F}{\sigma_F} \leq 2\sqrt{3}\left(p - \frac{1}{2}\right), \qquad (3.28)$$

which becomes an equality for the uniform distribution on $[\mu - \sqrt{3}\sigma, \mu + \sqrt{3}\sigma]$.

If $2/3 < p < 1$, then

$$\frac{F^{-1}(p) - \mu_F}{\sigma_F} \leq \left[\frac{9p - 5}{9(1-p)}\right]^{1/2}. \qquad (3.29)$$

This is an equality for the mixture of the Dirac distribution concentrated at $\mu - 3\sigma[(1-p)/(9p-5)]^{1/2}$ and the uniform one on $[\mu - 3\sigma[(1-p)/(9p-5)]^{1/2}, \mu + 3\sigma(3p-1)/[(1-p)(9p-5)]^{1/2}]$ with weights $3p-2$ and $3(1-p)$, respectively.

Proposition 2 (decreasing failure rate) *If $0 < p \leq 1 - e^{-1} \approx 0.63212$, then $F^{-1}(p) \leq \mu_F$.*

If $1 - e^{-1} < p \leq 1 - e^{-2} \approx 0.86466$, then bound

$$[F^{-1}(p) - \mu_F]/\sigma_F \leq -\ln(1-p) - 1, \qquad (3.30)$$

is attained by the exponential distribution function with location μ and scale σ.

If $1-e^{-2} < p < 1$, then for

$$\gamma = \gamma_V(p) = (1-p)e^2 \in (0,1),$$

we get

$$\frac{F^{-1}(p) - \mu_F}{\sigma_F} \le \left(\frac{2}{\gamma_V(p)} - 1\right)^{1/2}. \qquad (3.31)$$

This is an equality if F is the combination of an atom at $\mu - \sigma[\gamma/(2-\gamma)]^{1/2}$, and the exponential distribution with location $[\gamma(2-\gamma)]^{1/2}\mu - \gamma\sigma$ and scale σ, with respective probabilities $1-\gamma$ and γ.

We see that bounds (3.29) and (3.31) tend to infinity if $p \nearrow 1$, and the same holds generally for (3.26). It follows from the fact that for $b = W^{-1}(p)$ large enough, $\beta = \beta(b)$ satisfies

$$b = \frac{\int_\beta^d x(x-\beta)w(x)\,dx}{\int_\beta^d (x-\beta)w(x)\,dx} \qquad (3.32)$$

(cf. (3.9) for $\beta > a$). Therefore $b(\beta) \nearrow d$, as $\beta \nearrow d$, and the same holds for the inverse. Furthermore, $1 - W(\beta) \searrow 0$, whereas the nominator of $A^0_{\succeq_c W}$ remains bounded below from zero, as $b \nearrow d$.

Now we proceed to the class of distributions determined by relation $F \preceq_c W$. We look for the projection of an indicator function onto cone $\mathcal{C}^{\nearrow}_{\succeq_c W}$.

Lemma 6 *Under the hypotheses of Lemma 4, for every $g \in \mathcal{C}^{\nearrow}_{\succeq_c W}$ there exists $g_{\alpha\tilde{\beta}\gamma} \in \mathcal{C}^{\nearrow}_{\succeq_c W}$ defined as*

$$g_{\alpha\tilde{\beta}\gamma}(x) = \alpha[\min\{x,\gamma\} - \tilde{\beta}] \qquad (3.33)$$

for some $\alpha > 0$, and $a \le \tilde{\beta} \le b \le \gamma \le c$ with $\alpha(\gamma - \tilde{\beta}) \le M$ such that

$$\|g_{\alpha\tilde{\beta}\gamma} - h\| \le \|g - h\|.$$

PROOF. We start with eliminating all $g \in \mathcal{C}^{\nearrow}_{\succeq_c W}$ satisfying $g(c) \le 0$. Obviously, $g \le g_+ = \max\{g, 0\} \le h \ne g_+$. The first two relations show that g_+ provides a better approximation. The latter two imply that this can be further improved by adding a constant $C > 0$ such that

$$(g_+ + C, \mathbf{1}) = (h, \mathbf{1}).$$

We can repeat the truncation and translation operations several times until we eventually obtain a function positive at c.

Then we assume that $g(c) > 0$, and separately study three cases: $g(b) \leq 0$, $g(a) \leq 0 < g(b)$, and $g(a) > 0$. In the first one, we construct the line passing through $(b, 0)$ and tangent to the graph of g in $(b, c]$, and truncate it at level $g(c)$. The resulting function has the desired form, runs above g and below h in $[a, c)$, and the reversed relations hold in $[c, d)$. In the second case, we take the linear function secant to g at $\beta \in [a, b)$ satisfying $g(\beta) = 0$ and b. The truncation defined above provides the conclusions of the former case. Finally, if $g(a) > 0$, then it suffices to replace point $(\beta, 0)$ by $(a, 0)$ in the construction of the second case. ∎

The function (3.33) is a linear increasing function on the left which changes into a constant at point $\gamma \in [b, c]$. It is convenient to reparameterize (3.33) as

$$g_{\alpha\beta\gamma}(x) = \alpha[\min\{x, \gamma\} - \beta] + 1.$$

Then for fixed $\gamma \in [b, c]$ we minimize the squared distance

$$D(\alpha, \beta, \gamma) = \int_a^d [\alpha(\min\{x, \gamma\} - \beta) + 1 - h(x)]^2 w(x)\, dx \tag{3.34}$$

with respect to α and β. However, this is a simple matter, because (3.34) is a convex quadratic function in both arguments. Therefore we have

Lemma 7 *Under the hypotheses of Lemma 5, we have*

$$P^\wedge_{\preceq_c W} h(x) = \alpha(\gamma)[\min\{x, \gamma\} - \beta(\gamma)] + 1 \tag{3.35}$$

for some $\gamma \in [b, c]$ with

$$\beta(\gamma) = \int_a^d \min\{x, \gamma\} w(x)\, dx, \tag{3.36}$$

$$\alpha(\gamma) = \frac{\frac{\int_b^c \min\{x, \gamma\} w(x)\, dx}{\int_b^c w(x)\, dx} - \beta(\gamma)}{\int_a^d \min^2\{x, \gamma\} w(x)\, dx - \beta^2(\gamma)}. \tag{3.37}$$

In the description of quantile bounds for $F \preceq_c W$, we use moments of right truncated random variables with distribution function W (cf. (3.20) to (3.22)):

$$\bar{\mu}_W(\gamma) = E_W \min\{X, \gamma\} = \int_a^d \min\{x, \gamma\} w(x)\, dx, \tag{3.38}$$

$$\bar{\sigma}^2_W(\gamma) = \mathrm{Var}_W \min\{X, \gamma\}$$
$$= \int_a^d [\min\{x, \gamma\}]^2 w(x)\, dx - \bar{\mu}^2_W(\gamma), \tag{3.39}$$

$$\bar{\delta}_W(\gamma) = \frac{W^{-1}(p) - \bar{\mu}_W(\gamma)}{\bar{\sigma}_W(\gamma)}. \tag{3.40}$$

3.2 Distributions with Monotone Density and Failure Rate

Notice that (3.36) and (3.37) can be expressed as

$$\beta(\gamma) = \bar{\mu}_W(\gamma) \in [a_W, \gamma),$$
$$\alpha(\gamma) = \frac{E_W(\min\{X,\gamma\}|X > b) - \bar{\mu}_W(\gamma)}{\bar{\sigma}_W^2(\gamma)} > 0,$$

respectively. Therefore (3.35) is actually a nondecreasing concave function. In order to deduce the optimal bounds on quantiles, we do not need to calculate precisely parameter γ in Lemma 7, because we merely look for

$$g_* = \lim_{c \searrow b} P^{\nearrow}_{\preceq_c W} h \in C^{\nearrow}_{\preceq_c W}.$$

Note that (3.36) and (3.37) tend to $\beta(b) = \bar{\mu}_W(b)$, and

$$\alpha(b) = \frac{b - \beta(b)}{\int_a^d \min^2\{x,b\} w(x)\, dx - \beta^2(b)} = \frac{W^{-1}(p) - \bar{\mu}_W(b)}{\bar{\sigma}_W^2(b)}, \quad (3.41)$$

respectively, uniformly in all $\gamma \in [b, c]$, and therefore $P^{\nearrow}_{\preceq_c W} h$ tends to

$$g_*(x) = \alpha(b)[\min\{x,b\} - \beta(b)] + 1 \quad (3.42)$$

in the L^2-norm as $c \searrow b$. The norm

$$A^0_{\preceq_c W}(p) = \|g_* - 1\| = \alpha_*(b)\bar{\sigma}_W(b) = \bar{\delta}_W(b) \quad (3.43)$$

(cf. (3.38) to (3.40), and (3.41), and (3.42)) provides the optimal bound. Writing

$$\frac{F^{-1}W(x) - \mu_F}{\sigma_F} = \frac{g_*(x) - 1}{A^0_{\preceq_c W}(p)} = \frac{\min\{x,b\} - \bar{\mu}_W(b)}{\bar{\sigma}_W(b)}$$

(cf. (3.41) to (3.43)), we easily determine the distribution function that has a pth quantile at $A^0_{\preceq_c W}(p)$. In conclusion, we get the following result.

Theorem 5 ($F \preceq_c W$) *Then for arbitrary $0 < p < 1$ and $b = W^{-1}(p)$, we have*

$$\frac{F^{-1}(p) - \mu_F}{\sigma_F} \leq \bar{\delta}_W(b), \quad (3.44)$$

and the equality in (3.44) holds for

$$F(x) = \begin{cases} W\left(\bar{\mu}_W(b) + \bar{\sigma}_W(b)\frac{x-\mu}{\sigma}\right), & \text{if } \frac{x-\mu}{\sigma} < \bar{\delta}_W(b), \\ 1, & \text{if } \frac{x-\mu}{\sigma} \geq \bar{\delta}_W(b). \end{cases} \quad (3.45)$$

Formula (3.45) defines the distribution function of an affinely transformed random variable with distribution function W right truncated at its pth quantile. This becomes a life distribution if

$$\mu_F = \sigma_F[\bar{\mu}_W(b) - a_W]/\bar{\sigma}_W(b).$$

Bounds (3.44) continuously increase from 0 at $p = 0$ to $(d_W - \mu_W)/\sigma_W$ at $p = 1$, which is finite if W has a finite support. For large p, these are close to respective quantiles of parent W. However, they are positive for all quantiles in contrast to negative values of standardized quantiles of small orders for all $F \succeq_c W$ (including W itself).

Proposition 3 (increasing density) *For arbitrary $0 < p < 1$, we have*

$$\frac{F^{-1}(p) - \mu_F}{\sigma_F} \leq \left(\frac{3p}{4 - 3p}\right)^{1/2}. \tag{3.46}$$

Inequality (3.46) becomes an equality for a mixture of the uniform distribution on the interval $[\mu - \sigma\sqrt{3}(2-p)/[p(4-3p)]^{1/2}, \mu + \sigma[3p/(4-3p)]^{1/2}]$ and the degenerate measure concentrated at $\mu + \sigma[3p/(4-3p)]^{1/2}$ with probabilities p and $1-p$, respectively.

Proposition 4 (increasing failure rate) *For arbitrary $0 < p < 1$, we have*

$$\frac{F^{-1}(p) - \mu_F}{\sigma_F} \leq \frac{-\ln(1-p) - p}{[p(2-p) + 2(1-p)\ln(1-p)]^{1/2}}. \tag{3.47}$$

The equality in (3.47) is attained by the distribution of the random variable

$$Y = \mu + \sigma \frac{\min\{X, -\ln(1-p) - p\}}{[p(2-p) + 2(1-p)\ln(1-p)]^{1/2}}$$

for a standard exponentially distributed X.

3.3 Distributions with Monotone Density and Failure Rate on the Average

First we consider quantile bounds for $F \succeq_* W$ for which the projection of $h = M\mathbf{1}_{[b,c)}$ onto the family of nondecreasing starshaped functions is needed.

Lemma 8 *For h defined in Lemma 4, and for every $g \in \mathcal{C}^{\nearrow}_{\succeq_*W}$ there exists $g_{\alpha\beta} \in \mathcal{C}^{\nearrow}_{\succeq_*W}$ defined as*

$$g_{\alpha\beta}(x) = \alpha(x - a)\mathbf{1}_{[b,d)}(x) + \beta \tag{3.48}$$

for some $\alpha, \beta \geq 0$, $\alpha(b-a) + \beta < M$, such that

$$\|g_{\alpha\beta} - h\| \leq \|g - h\|.$$

3.3 Distributions with Monotone Density and Failure Rate on the Average 45

Candidates for projection $P^{\nearrow}_{\succeq_* W} h$ have a constant value $\beta \in [0, h(b))$ in $[a, b)$, a jump at b to $g_{\alpha\beta}(b) \stackrel{<}{} h(b)$, and a linear part in $[b, d)$ that can be extended to the left so as to pass through $(a, \beta) = (a, g_{\alpha\beta}(a))$. A constant projection β is also possible.

PROOF OF LEMMA 8. The first steps of reasoning are similar to those of Lemma 4. If $g \in \mathcal{C}^{\nearrow}_{\succeq_* W}$, so does g_+. Moreover, the latter lies closer to h. If $g(c) \leq 0$, then a zero function better approximates h, and can be further improved by constant $1 = (h, 1)$. If

$$g(b) \geq h(b) = \max_{a \leq x < d} h(x),$$

then a downward translation of g gives a better approximation. Therefore we can confine ourselves to functions $g \in \mathcal{C}^{\nearrow}_{\succeq_* W}$ such that $0 \leq g(a) \leq g(b) < h(b)$ and $g(c) > 0$. Let $\gamma \in (b, c]$ be the smallest point at which g exceeds h. Put

$$g_\gamma(x) = g(a) + \frac{g(\gamma) - g(a)}{\gamma - a}(x - a)\mathbf{1}_{[b,d)}(x). \tag{3.49}$$

This is the closest function to h in $[a, b)$ among all nondecreasing starshaped functions starting from level $g(a) \geq 0$. Moreover, $g \in \mathcal{C}^{\nearrow}_{\succeq_* W}$ implies that

$$g(x) \leq g_\gamma(x) \leq h(x), \quad \text{if} \quad x \in [b, \gamma),$$
$$g(x) \geq g_\gamma(x) \geq h(x), \quad \text{if} \quad x \in [\gamma, d).$$

This gives the desired conclusion, because (3.49) can be easily reparameterized as (3.48). ■

Lemma 9 *Let assumptions of Lemma 4 and (3.7) hold. If*

$$\frac{\int_b^c (x-a)w(x)\,dx}{\int_b^c w(x)\,dx} \leq \int_b^d (x-a)w(x)\,dx, \tag{3.50}$$

then $P^{\nearrow}_{\succeq_* W} h = 1$. *Otherwise*

$$P^{\nearrow}_{\succeq_* W} h(x) = \alpha_*(x-a)\mathbf{1}_{[b,d)}(x) + \beta_* \tag{3.51}$$

for $\alpha_* > 0$, $0 \leq \beta_* < 1$ *defined by*

$$\alpha_* = \frac{\frac{\int_b^c (x-a)w(x)\,dx}{\int_b^c w(x)\,dx} - \int_b^d (x-a)w(x)\,dx}{\int_b^d (x-a)^2 w(x)\,dx - \left[\int_b^d (x-a)w(x)\,dx\right]^2}, \tag{3.52}$$

$$\beta_* = \frac{\int_b^d (x-a)^2 w(x)\,dx - \frac{\int_b^c (x-a)w(x)\,dx}{\int_b^c w(x)\,dx}\int_b^d (x-a)w(x)\,dx}{\int_b^d (x-a)^2 w(x)\,dx - \left[\int_b^d (x-a)w(x)\,dx\right]^2}. \tag{3.53}$$

PROOF. By Lemma 8, it suffices to minimize the function

$$D(\alpha, \beta) = \|g_{\alpha\beta} - h\|^2$$

$$= \int_a^d \left[\beta + \alpha(x-a)\mathbf{1}_{[b,d]}(x) - \frac{\mathbf{1}_{[b,c)}(x)}{\int_b^c w(x)\,dx} \right]^2 w(x)\,dx$$

$$= \beta^2 - 2\beta \cdot \frac{1}{\int_b^c w(x)\,dx} + \alpha^2 \int_b^d (x-a)^2 w(x)\,dx$$

$$- 2\alpha \left[\frac{\int_b^c (x-a)w(x)\,dx}{\int_b^c w(x)\,dx} - \beta \int_b^d (x-a)w(x)\,dx \right] \quad (3.54)$$

with respect to parameters $\alpha \geq 0$ and $0 \leq \beta < 1/[\int_b^c w(x)\,dx]$. Fixing α, we minimize (3.54) at

$$\beta_*(\alpha) = 1 - \alpha \int_b^d (x-a)w(x)\,dx. \quad (3.55)$$

Plugging it into (3.54), we obtain

$$D(\alpha, \beta_*(\alpha)) = \alpha^2 \left\{ \int_b^d (x-a)^2 w(x)\,dx - \left[\int_b^d (x-a)w(x)\,dx \right]^2 \right\}$$

$$- 2\alpha \left[\frac{\int_b^c (x-a)w(x)\,dx}{\int_b^c w(x)\,dx} - \int_b^d (x-a)w(x)\,dx \right]$$

$$+ \frac{1}{\int_b^c w(x)\,dx} - 1. \quad (3.56)$$

This is a quadratic function of α, with a positive coefficient of the quadratic term. Under the nonnegativity condition, (3.56) is minimized at the maximum of 0 and (3.52). If (3.50) holds, then the optimal value is $\alpha = 0$, and, by (3.55), $\beta = 1$. This proves the first statement. Otherwise (3.56) is minimized by (3.52), for which (3.55) coincides with (3.53). ∎

Notice that the denominator of (3.52) is positive by the Schwarz inequality with $\int_b^d w(x)\,dx < 1$, and so is the numerator if (3.50) does not hold. Therefore (3.51) is actually a nondecreasing starshaped function. Observe finally that (3.53) is positive, because its numerator, divided by $1 - W(b)$, can be interpreted as

$$E_W((X-a)^2|X>b) - E_W(X-a|b<X<c)E_W(X-a|X>b)$$
$$\geq \text{Var}_W(X|X>b) > 0.$$

Moreover, by (3.55), $\beta < 1 < 1/[\int_b^c w(x)\,dx]$, and so (3.53) satisfies both constraints deduced from the geometric arguments presented in Lemma 8.

3.3 Distributions with Monotone Density and Failure Rate on the Average

By the arguments of Section 2.3,

$$P^0_{\succeq_* W}(h-1) = P^\nearrow_{\succeq_* W} h - 1.$$

This is 0 if (3.50) holds. Otherwise, we have

$$\|P^0_{\succeq_* W} h - 1\| = \frac{\frac{\int_b^c (x-a) w(x)\,dx}{\int_b^c w(x)\,dx} - \int_b^d (x-a) w(x)\,dx}{\left\{\int_b^d (x-a)^2 w(x)\,dx - \left[\int_b^d (x-a) w(x)\,dx\right]^2\right\}^{1/2}}, \quad (3.57)$$

$$\frac{P^0_{\succeq_* W} h(x) - 1}{\|P^0_{\succeq_* W} h - 1\|} = \frac{(x-a) \mathbf{1}_{[b,d)}(x) - \int_b^d (y-a) w(y)\,dy}{\left\{\int_b^d (y-a)^2 w(y)\,dy - \left[\int_b^d (y-a) w(y)\,dy\right]^2\right\}^{1/2}}. \quad (3.58)$$

Letting $c \searrow b$, and using the notation of Theorem 4, we state its analogue for $F \succeq_* W$.

Theorem 6 $(F \succeq_* W)$ *Using (3.20) and (3.21) with $\beta = b = W^{-1}(p)$, write*

$$\eta = \eta_W(a,b) = E_W[(X-a)\mathbf{1}_{[b,d)}(X)]$$
$$= \int_b^d (x-a) w(x)\,dx = (1-p)\mu_W(b), \quad (3.59)$$

$$\vartheta^2 = \vartheta_W^2(a,b) = \mathrm{Var}_W[(X-a)\mathbf{1}_{[b,d)}(X)]$$
$$= \int_b^d (x-a)^2 w(x)\,dx - \left[\int_b^d (x-a) w(x)\,dx\right]^2. \quad (3.60)$$

If $\eta_W(a,b) \geq b-a$, then $F^{-1}(p) \leq \mu_F$.
Otherwise

$$\frac{F^{-1}(p) - \mu_F}{\sigma_F} \leq A^0_{\succeq_* W}(p) = \frac{b - a - \eta_W(a,b)}{\vartheta_W(a,b)}, \quad (3.61)$$

and bound (3.61) is attained by

$$F(x) = \begin{cases} 0, & \text{if } \frac{x-\mu}{\sigma} < -\frac{\eta}{\vartheta}, \\ p, & \text{if } -\frac{\eta}{\vartheta} \leq \frac{x-\mu}{\sigma} \leq \frac{b-a-\eta}{\vartheta}, \\ W\left(b - \eta + \vartheta \frac{x-\mu}{\sigma}\right), & \text{if } \frac{x-\mu}{\sigma} \geq \frac{b-a-\eta}{\vartheta}. \end{cases} \quad (3.62)$$

Obviously, $A^0_{\succeq_* W}(p) \geq A^0_{\succeq_c W}(p)$, because $F \succeq_c W$ implies $F \succeq_* W$. It follows that (3.61) tends to infinity as $p \nearrow 1$. The main difference between (3.27) and (3.62) is that the atom of the latter is distant from the smooth part. Special cases of distributions dominating the uniform and exponential ones in the star order are presented in Propositions 5 and 6.

3. Quantiles

Proposition 5 (decreasing density on the average) *If we have $p \leq \sqrt{2} - 1 \approx 0.41421$, then $F^{-1}(p) \leq \mu_F$.*
If $p > \sqrt{2} - 1$, then

$$\frac{F^{-1}(p) - \mu_F}{\sigma_F} \leq \frac{\sqrt{3}(p^2 + 2p - 1)}{\theta_U(p)} \qquad (3.63)$$

with

$$\theta_U^2(p) = 12\vartheta_U^2(0, p) = 1 + 6p^2 - 4p^3 - 3p^4. \qquad (3.64)$$

Relation (3.63) becomes an equality for the mixture of an atom at $\mu - \sqrt{3}\sigma(1-p^2)/\theta_U(p)$ with probability p, and the uniform distribution on $[\mu + \sqrt{3}\sigma(p^2 + 2p - 1)/\theta_U(p), \mu + \sqrt{3}\sigma(1+p^2)/\theta_U(p)]$ with probability $1 - p$.

Proposition 6 (decreasing failure rate on the average) *Let us denote by $p_0 \approx 0.553567$ the unique zero of the strictly increasing function*

$$\nu_V(p) = p[1 - \ln(1-p)] - 1, \quad 0 < p < 1. \qquad (3.65)$$

If $p \leq p_0$, then $F^{-1}(p) \leq \mu_F$.
Otherwise

$$\frac{F^{-1}(p) - \mu_F}{\sigma_F} \leq \frac{\nu_V(p)}{\theta_V(p)} \qquad (3.66)$$

for

$$\theta_V^2(p) = \vartheta_V^2(0, -\ln(1-p)) = (1-p)p[1 - \ln(1-p)]^2 + 1 - p. \qquad (3.67)$$

Bound (3.66) is sharp. This is attained by F being a combination of the jump distribution at $\mu - \sigma(1-p)[1 - \ln(1-p)]/\theta_V(p)$ and the exponential distribution with location $\mu + \sigma\nu_V(p)/\theta_V(p)$ and scale $\sigma(1-p)/\theta_V(p)$.

Theorem 7 asserts that general bound (3.2) cannot be improved in classes of distributions defined by $F \preceq_* W$. Moreover, we conclude analogous results for all F preceding a fixed W in partial order more general than the star one, for example the superadditive and Laplace transform orders (see, e.g., Shaked and Shantikumar [96]).

Theorem 7 ($F \preceq_* W$) *If density function $w(x)$ is bounded on a neighborhood of a_W, then for arbitrary $0 < p < 1$ there exist sequences of absolutely continuous distribution functions $F_k \preceq_* W$ with arbitrary common mean μ and variance σ^2 such that*

$$\lim_{k \to \infty} \frac{F_k^{-1}(p) - \mu}{\sigma} = \left(\frac{p}{1-p}\right)^{1/2}.$$

In particular, the statement holds for distributions with increasing density and failure rate on the average.

3.3 Distributions with Monotone Density and Failure Rate on the Average

The claim can be justified by some informal arguments. To fix the ideas, we consider the approximation of two-point distribution F defined in Theorem 2 by $F_k \preceq_* U$, that is, ones with nondecreasing $F_k(x)/(x - a_k)$ on respective supports $[a_k, d_k)$. Note that F is constant beside of its two jumps, and the starshaped F_k have to increase at rate $1/(x - a_k)$ at least on their supports. However, letting $a_k \to -\infty$, we relax conditions on the increase rate of F_k itself. Accordingly, such F_k can be starshaped and approximate well any two-point distribution at and beside jump points. The formal proof of Theorem 7 is constructive. It is also possible to provide special constructions of sequences satisfying the statement for specific W with unbounded densities about respective left support ends.

PROOF OF THEOREM 7. Due to Theorem 2 the general bound (3.31) is attained by the quantile functions

$$F^{-1}(x) = \begin{cases} \mu_F - \sigma_F \sqrt{\frac{1-p}{p}} & \text{if } 0 \leq x < p, \\ \mu_F + \sigma_F \sqrt{\frac{p}{1-p}} & \text{if } p \leq x < 1. \end{cases}$$

Under a change of variables, we have

$$\frac{F^{-1}W(x) - \mu_F}{\sigma_F} = g_p(x) = \begin{cases} -\sqrt{\frac{1-p}{p}}, & \text{if } a \leq x < b = W^{-1}(p), \\ \sqrt{\frac{p}{1-p}}, & \text{if } b \leq x < d. \end{cases} \tag{3.68}$$

Our proof consists in constructing a sequence $g_k \in \mathcal{C}^{\nearrow}_{\preceq_* W}$, $k \to \infty$, that converges to (3.68) in $L^2([a_W, d_W), w(x)dx)$. For sufficiently large k, we define piecewise linear continuous nondecreasing starshaped functions

$$\tilde{g}_k(x) = \begin{cases} k^4(x-a) - k - \sqrt{\frac{1-p}{p}}, & \text{if } a \leq x \leq a + k^{-3}, \\ -\sqrt{\frac{1-p}{p}}, & \text{if } a + k^{-3} \leq x \leq b, \\ k\frac{x-a}{b-a} - k - \sqrt{\frac{1-p}{p}}, & \text{if } b \leq x \leq b + \frac{b-a}{k\sqrt{p(1-p)}}, \\ \sqrt{\frac{p}{1-p}}, & \text{if } b + \frac{b-a}{k\sqrt{p(1-p)}} \leq x < d. \end{cases}$$

Observe that

$$\|\tilde{g}_k - g_p\|^2 = \int_a^{a+k^{-3}} [k^4(x-a) - k]^2 w(x) \, dx$$
$$+ \int_b^{b + \frac{b-a}{k\sqrt{p(1-p)}}} \left[k\frac{x-a}{b-a} - k - \frac{1}{\sqrt{p(1-p)}}\right]^2 w(x) \, dx$$
$$\leq \sup_{a \leq x \leq a + k^{-3}} \frac{w(x)}{3k} + \frac{W\left(b + \frac{b-a}{k\sqrt{p(1-p)}}\right) - W(b)}{p(1-p)} \to 0$$

as $k \to \infty$. The same holds for
$$g_k(x) = \frac{\tilde{g}_k(x) - (\tilde{g}_k, \mathbf{1})}{||\tilde{g}_k - (\tilde{g}_k, \mathbf{1})||},$$
because
$$(\tilde{g}_k, \mathbf{1}) \to (\tilde{g}_p, \mathbf{1}) = 0,$$
and
$$||\tilde{g}_k - (\tilde{g}_k, \mathbf{1})|| \to ||g_p|| = 1.$$
Relations
$$[F_k^{-1} W(x) - \mu_F]/\sigma_F = g_k(x)$$
define a sequence of $F_k \preceq_* W$ with common mean μ and variance σ^2 for which $F_k^{-1}(p) \nearrow \sqrt{p/(1-p)}$. ∎

3.4 Symmetric Unimodal Distributions

Let W be an absolutely continuous distribution function of a symmetric random variable. We are interested in evaluating $[F^{-1}(p) - \mu_F]/\sigma_F$ for all symmetric distributions satisfying $F \succeq_s W$. A dual problem is to project

$$h^s(x) = \begin{cases} -\dfrac{\mathbf{1}_{[W^{-1}(1-q),W^{-1}(1-p))}(x)}{2(q-p)} & \text{if } p < 1/2, \\ +\dfrac{\mathbf{1}_{[W^{-1}(p),W^{-1}(q))}(x)}{2(q-p)} & \text{if } p \geq 1/2, \end{cases} \quad (3.69)$$

for $q \searrow p$ onto $\mathcal{C}^s_{\succeq_c W}$ as defined in (2.60). If $p < 1/2$, then $h^s \leq 0$. Its projection onto the family of nonnegative functions $\mathcal{C}^+ \supset \mathcal{C}^s_{\succeq_c W}$ is $P^+ h^s = 0$ which actually belongs to $\mathcal{C}^s_{\succeq_c W}$, and so $P^s_{\succeq_c W} h^s = 0$. This sequence of elaborated arguments leads us to the trivial conclusion

$$F^{-1}(p) \leq F^{-1}\left(\frac{1}{2}\right) = \mu_F, \quad p < \frac{1}{2}.$$

Otherwise we have

Lemma 10 *Let*

$$h^s = M \mathbf{1}_{[b,c)} \in L^2([\mu_W, d_W), 2w(x)dx)$$

for $\mu_W < b < c \leq d_W$ and $M > 0$. Then for every $g \in \mathcal{C}^s_{\succeq_c W}$ there exists $g_{\alpha\beta} \in \mathcal{C}^s_{\succeq_c W}$ defined as

$$g_{\alpha\beta}(x) = \alpha(x - \beta)_+$$

for some $\alpha \geq 0$, $\mu_W \leq \beta \leq b$, $\alpha(b - \beta) < M$ such that

$$||g_{\alpha\beta} - h^s|| \leq ||g - h^s||.$$

3.4 Symmetric Unimodal Distributions

The only difference between the statements of Lemmas 4 and 10 is that there is a lower constraint on parameter β in the latter. If b is close to μ_W, the constraint plays a significant role leading to conclusions essentially different from those of Lemma 5.

Lemma 11 *If for $\mu = \mu_W$, we have*

$$\frac{\int_b^c (x-\mu)w(x)\,dx}{\int_b^c w(x)\,dx} \leq \frac{\int_\mu^d (x-\mu)^2 w(x)\,dx}{\int_\mu^d (x-\mu)w(x)\,dx}, \tag{3.70}$$

then

$$P_{\succeq_c W}^s h^s(x) = \frac{\int_b^c (y-\mu)w(y)\,dy}{2\int_b^c w(y)\,dy} \frac{x-\mu}{\int_\mu^d (y-\mu)^2 w(y)\,dy}. \tag{3.71}$$

Otherwise there exists a unique $\mu < \beta_* < b$ that solves

$$\frac{\int_b^c (x-\beta)w(x)\,dx}{\int_b^c w(x)\,dx} = \frac{\int_\beta^d (x-\beta)^2 w(x)\,dx}{\int_\beta^d (x-\beta)w(x)\,dx} \tag{3.72}$$

(cf. (3.9)) such that

$$P_{\succeq_c W}^s h^s(x) = \frac{(x-\beta_*)_+}{2\int_{\beta_*}^d (y-\beta_*)w(y)\,dy}. \tag{3.73}$$

Under (3.70), we have

$$\|P_{\succeq_c W}^s h^s\|^2 = \frac{\left[\int_b^c (x-\mu)w(x)\,dx\right]^2}{2\left[\int_b^c w(x)\,dx\right]^2 \int_\mu^d (x-\mu)^2 w(x)\,dx}, \tag{3.74}$$

$$\frac{P_{\succeq_c W}^s h^s(x)}{\|P_{\succeq_c W}^s h^s\|} = \frac{x-\mu}{\left[2\int_\mu^d (y-\mu)^2 w(y)\,dy\right]^{1/2}}. \tag{3.75}$$

Otherwise

$$\|P_{\succeq_c W}^s h^s\|^2 = \frac{\int_{\beta_*}^d (x-\beta_*)^2 w(x)\,dx}{2\left[\int_{\beta_*}^d (x-\beta_*)w(x)\,dx\right]^2}, \tag{3.76}$$

$$\frac{P_{\succeq_c W}^s h^s(x)}{\|P_{\succeq_c W}^s h^s\|} = \frac{(x-\beta_*)_+}{\left[2\int_{\beta_*}^d (y-\beta_*)^2 w(y)\,dy\right]^{1/2}}. \tag{3.77}$$

Letting $c \searrow b$, recalling (3.20) through (3.22), and introducing

$$\varsigma_W = \mathrm{E}_W |X - \mu_W| = 2\int_\mu^d (x-\mu)w(x)\,dx \tag{3.78}$$

(the latter under the symmetry assumption), we obtain conclusions similar to ones obtained in Theorem 4.

Theorem 8 ($F \succeq_s W$) If $p \leq 1/2$ (i.e., $W^{-1}(p) \leq \mu_W$), then we have $F^{-1}(p) \leq \mu_F$.

If $\mu_W < W^{-1}(p) \leq \mu_W + \sigma_W^2/\varsigma_W$, then (3.23) holds, with the equality attained by (3.24).

Finally, if $W^{-1}(p) > \mu_W + \sigma_W^2/\varsigma_W$, then there exists a unique $\beta_* = \beta_*(p) \in (\mu_W, W^{-1}(p))$ solving Equation (3.25) such that

$$\frac{F^{-1}(p) - \mu_F}{\sigma_F} \leq A = A^s_{\succeq_s W}(p) = \left\{ \frac{\delta_W^2(\beta_*) + 1}{2[1 - W(\beta_*)]} \right\}^{1/2}. \qquad (3.79)$$

The equality in (3.79) holds for

$$F(x) = \begin{cases} 1 - W\left(\beta_* - 2[\mu_W(\beta_*) - \beta_*]A\frac{x-\mu}{\sigma}\right), & \text{if } \frac{x-\mu}{\sigma} < 0, \\ W\left(\beta_* + 2[\mu_W(\beta_*) - \beta_*]A\frac{x-\mu}{\sigma}\right), & \text{if } \frac{x-\mu}{\sigma} \geq 0. \end{cases} \qquad (3.80)$$

Symmetric distribution function (3.80) has a jump of height $2W(\beta_*) - 1 < 2p - 1$ at its center of symmetry. If $W^{-1}(p) = \mu_W + \sigma_W^2/\varsigma_W$, then μ_W solves (3.25), bounds (3.23) and (3.79) are equal, and (3.80) coincides with (3.24). Putting $W = U$, we get

Proposition 7 (symmetric unimodal distributions) If $p \leq 1/2$, then $F^{-1}(p) \leq \mu_F$.

If $1/2 \leq p \leq 5/6$, then

$$\frac{F^{-1}(p) - \mu_F}{\sigma_F} \leq 2\sqrt{3}\left(p - \frac{1}{2}\right), \qquad (3.81)$$

which becomes the equality for the uniform distribution on interval $[\mu - \sqrt{3}\sigma, \mu + \sqrt{3}\sigma]$.

If $5/6 \leq p < 1$, then

$$\frac{F^{-1}(p) - \mu_F}{\sigma_F} \leq \frac{1}{3}\left(\frac{2}{1-p}\right)^{1/2}. \qquad (3.82)$$

The bound becomes the equality for a mixture of the atom at μ with probability $6p - 5$, and uniform distribution on interval $[\mu - \sigma/\sqrt{2(1-p)}, \mu + \sigma/\sqrt{2(1-p)}]$ with probability $6(1-p)$.

Observe that (3.82) is 1.5 times less than respective bound (3.5) for symmetric distributions.

In Table 3.1, we numerically compare bounds on standardized quantiles $[F^{-1}(p) - \mu_F]/\sigma_F$ of orders $p = 0.05(0.05)0.95$ for nine families of distributions: general (G), symmetric (S), and symmetric unimodal (SUN) ones, distributions with decreasing density (DD) and failure rate (DFR), and those on the average (DDA and DFRA, respectively), and with increasing density (ID) and failure rate (IFR). Bounds for distributions with increasing density and failure rate on the average are identical with the general

TABLE 3.1. Sharp uniform mean-variance bounds on quantiles for various families of distributions.

p	G	S	SUN	DD	DFR	DDA	DFRA	ID	IFR
0.05	0.22942	0	0	0	0	0	0	0.19739	0.19782
0.10	0.33333	0	0	0	0	0	0	0.28475	0.28601
0.15	0.42008	0	0	0	0	0	0	0.35603	0.35874
0.20	0.5	0	0	0	0	0	0	0.42008	0.42465
0.25	0.57735	0	0	0	0	0	0	0.48038	0.48741
0.30	0.65465	0	0	0	0	0	0	0.53882	0.54905
0.35	0.73380	0	0	0	0	0	0	0.59660	0.61097
0.40	0.81650	0	0	0	0	0.13511	0	0.65465	0.67435
0.45	0.90453	0	0	0	0	0.32163	0	0.71375	0.74030
0.50	1	1	0	0	0	0.50913	0	0.77460	0.81005
0.55	1.10554	1.05409	0.17321	0.17321	0	0.70235	0.15541	0.83793	0.88503
0.60	1.22474	1.11803	0.34641	0.34641	0.04982	0.90763	0.33725	0.90453	0.96708
0.65	1.36277	1.19523	0.51962	0.51962	0.20397	1.13406	0.53901	0.97531	1.05869
0.70	1.52753	1.29099	0.69282	0.69389	0.38629	1.39582	0.77040	1.05131	1.16351
0.75	1.73205	1.41421	0.86603	0.88192	0.60944	1.71781	1.05060	1.13389	1.28721
0.80	2	1.58114	1.03923	1.10554	0.89712	2.15063	1.42081	1.22474	1.43941
0.85	2.38048	1.82574	1.21716	1.40106	1.30641	2.82311	1.98883	1.32613	1.63860
0.90	3	2.23607	1.49071	1.85592	1.30641	2.82311	1.98883	1.44115	1.92754
0.95	4.35890	3.16228	2.10819	2.80872	2.10081	4.24076	3.18532	1.57425	2.44875

ones. Certainly, the general bounds are merely valid for wider classes of life distributions (e.g., NBU, HNBUE and $\mathcal{L}$-class defined in Klefsjö [46]) which are generated by more general orders than the star order. On the other hand, there are still interesting open problems of accurate bounds on quantiles of distributions succeeding a given one in these general orders.

3.5 Open Problems

1. What are the bounds on standardized quantiles in families of distributions defined by relation $F \preceq_s W$ for a fixed symmetric distribution function W? Of special interest are the symmetric U-shaped distributions, alternatively defined by $F \preceq_s U$.

2. Mean-variance bounds on quantiles of distributions determined by $F \preceq_* W$ coincide with general bounds under very mild conditions on fixed W. The reason is that without disturbing values of mean, variance, and quantile of a given order we are able to choose elements of the class with an arbitrarily remote left endpoint of support which approximate two-point distributions attaining the general bounds. This is impossible if we fix the left endpoint ($a_F = 0$, say) and the second raw moment m_F^2. Then we have the problem of evaluating $F^{-1}(p)/m_F$ for all life distributions determined by $F \preceq_* W$ with a nontrivial solution. Also, still unknown are the bounds on $F^{-1}(p)/m_F$ for F coming from the classes of

   (a) general distributions,
   (b) distributions succeeding (preceding) fixed W in the convex order, and
   (c) distributions succeeding fixed W in the star order.

3. Quantile differences $F^{-1}(q) - F^{-1}(p)$, $0 < p < q < 1$, are measures of population dispersion less sensitive to errors in statistical sampling than the standard deviation σ_F. Formula (3.3) evaluates the ratio $[F^{-1}(q) - F^{-1}(p)]/\sigma_F$ for arbitrary F. What are the respective bounds if F belongs to the restricted families of

   (a) symmetric distributions,
   (b) distributions determined by relations in the convex order,
   (c) distributions determined by relations in the star order, and
   (d) distributions determined by relations in the s-order?

   Of special interest, especially in symmetric populations, are quantile differences for symmetric pairs p and $q = 1 - p$.

4
Order Statistics of Independent Samples

Bounds for expectations of order statistics from general i.i.d. samples, due to Moriguti [58], are presented in Section 4.1. Bounds of Sections 4.2 and 4.4 for populations with decreasing density and failure rate, and symmetric unimodal ones, were obtained by Gajek and Rychlik [33]. Results of Section 4.3 for restricted families determined by the star order come from Rychlik [89]. Section 4.5, partially based on Okolewski and Rychlik [65], is devoted to the study of quantile estimation bias in various nonparametric families of distributions.

We do not discuss here bounds for the sample maximum and range from discrete populations obtained by López-Blázquez [50, 51]. We also merely mention sharp inequalities due to Papadatos [69] for the expectations of order statistics and their differences of nonnegative samples expressed in terms of the population mean. Blom [17] and van Zwet [100] developed another method of evaluating the expectations of order statistics from restricted families of distributions defined by means of stochastic orders using quantiles of the parent distribution. The method is based on the Jensen inequality. Papadatos [67, 68] derived some attainable bounds on the variance of order statistics from general and symmetric distributions, respectively, measured in the standard deviation units of the parent distribution. Analogous evaluations for covariances of order statistics are known, but only small samples of size $n = 2$ and 3 were treated (see, e.g., Ma [52] and Papadatos [70]). Bounds and approximations for moments of order statistics were reviewed in David [22, Chapter 4], Arnold and Balakrishnan [5, Chapters 4, 5], and Rychlik [84].

4.1 General and Symmetric Distributions

The problem of calculating mean-variance bounds on expectations of order statistics lies in determining projections of density functions (2.22) onto the convex cone (2.46). Observe that

$$f_{n:n}(x) = nx^{n-1}, \quad 0 < x < 1, \tag{4.1}$$

is actually nondecreasing, and so $P^\nearrow f_{n:n} = f_{n:n}$, and $P^0(f_{n:n} - 1) = f_{n:n} - 1$. Therefore

$$\frac{E_F X_{n:n} - \mu_F}{\sigma_F} \leq \|f_{n:n} - 1\| = \frac{n-1}{(2n-1)^{1/2}} \sim (n/2)^{1/2} \tag{4.2}$$

with

$$F(x) = \left[\frac{1}{n}\left(1 + \frac{n-1}{(2n-1)^{1/2}}\frac{x-\mu}{\sigma}\right)\right]^{1/(n-1)},$$

$$-\frac{(2n-1)^{1/2}}{n-1} \leq \frac{x-\mu}{\sigma} \leq (2n-1)^{1/2}, \tag{4.3}$$

attaining the bound. Distribution (4.3) is a location-scale transformation of a power distribution, and this is uniform for $n = 2$. The result was published independently by Gumbel [36] and Hartley and David [38]. More generally, the respective bound for L-statistics

$$E_F \sum_{j=1}^n c_j(X_{j:n} - \mu_F)/\sigma_F \leq \left\|\sum_{j=1}^n c_j(f_{j:n} - 1)\right\| \tag{4.4}$$

is sharp if the argument of the norm is nondecreasing. By differentiation,

$$\sum_{j=1}^n c_j f'_{j:n}(x) = n \sum_{j=1}^{n-1}(c_{j+1} - c_j)f_{j:n-1}(x), \tag{4.5}$$

we conclude that nondecrease of the sequence of coefficients is a sufficient condition. This was implicitly exploited by Plackett [72] and Nagaraja [60] in calculating optimal bounds for the sample range $X_{n:n} - X_{1:n}$ and *selection differentials* $1/k \sum_{j=n+1-k}^n X_{j:n}$, respectively.

In the case of general bounds, the Moriguti [58] projection method should be used (cf. Rychlik [84, Theorem 7, p. 121]). Here we confine ourselves to single order statistics. Observe first that $E_F X_{1:n} \leq \mu_F$ is implied by relation $X_{1:n} \leq X_1$. Calculating bounds for nonextreme order statistics $X_{j:n}$, $2 \leq j \leq n-1$, we refer to Moriguti [58, Example 2, p. 111]. We first aim at determining $P^\nearrow f_{j:n}$. Note that $f_{j:n}(x) > 0$ for $0 < x < 1$, and increasing-decreasing with the maximum at $(j-1)/(n-1)$. Therefore $F_{j:n}$

is strictly increasing on $[0,1]$, convex on $[0,(j-1)/(n-1)]$, and concave on $[(j-1)/(n-1),1]$. The slopes of tangent lines

$$l_\alpha(x) = f_{j:n}(\alpha)(x-\alpha) + F_{j:n}(\alpha)$$

continuously increase for $\alpha \in [0,(j-1)/(n-1)]$ and so do $l_\alpha(1)$, ranging from 0 to $l_{(j-1)/(n-1)}(1) > F_{j:n}(1) = 1$. The line l_{α_*} for some $\alpha_* \in (0,(j-1)/(n-1))$ such that $l_{\alpha_*}(1) = 1$ becomes a part of the lower convex envelope of $F_{j:n}$ on $[\alpha_*,1]$, and the remaining part coincides with $F_{j:n}$. Therefore, the greatest convex minorant $\bar{F}_{j:n}$ of $F_{j:n}$ has form

$$\bar{F}_{j:n}(x) = \begin{cases} F_{j:n}(x), & \text{if } 0 \le x \le \alpha_*, \\ f_{j:n}(\alpha_*)(x-1) + 1, & \text{if } \alpha_* \le x < 1, \end{cases} \quad (4.6)$$

for a unique $\alpha_* = \alpha_*(j,n) \in (0,(j-1)/(n-1))$ satisfying

$$(1-\alpha_*)f_{j:n}(\alpha_*) = 1 - F_{j:n}(\alpha_*). \quad (4.7)$$

The derivative of (4.6) is

$$P^\nearrow f_{j:n}(x) = \bar{f}_{j:n}(x) = f_{j:n}(\min\{x,\alpha_*\}). \quad (4.8)$$

Using

$$f_{i:m}(x)f_{j:n}(x) = n\frac{\binom{i+j-2}{i-1}\binom{m+n-i-j}{m-i}}{\binom{m+n-1}{m}} f_{i+j-1:m+n-1}(x), \quad (4.9)$$

we calculate

$$\begin{aligned}\|\bar{f}_{j:n}\|^2 &= \int_0^{\alpha_*} f_{j:n}^2(x)\,dx + (1-\alpha_*)f_{j:n}^2(\alpha_*) \\ &= n\frac{\binom{2j-2}{j-1}\binom{2n-2j}{n-j}}{\binom{2n-1}{n}} F_{2j-1:2n-1}(\alpha_*) \\ &\quad + (1-\alpha_*)f_{j:n}^2(\alpha_*). \end{aligned} \quad (4.10)$$

We have thus proven

Theorem 9 (general distributions) *For arbitrary F and $j = 1$, we have $\mathrm{E}_F X_{1:n} \le \mu_F$, which is attained by the atom measure at μ_F.*

For $2 \le j \le n-1$, inequality

$$[\mathrm{E}_F X_{j:n} - \mu_F]/\sigma_F \le B = B^0(j,n) = (\|\bar{f}_{j:n}\|^2 - 1)^{1/2} \quad (4.11)$$

is sharp (see (4.7) and (4.10)), and this is attained by the distribution function

$$F(x) = \begin{cases} 0, & \text{if } \frac{x-\mu}{\sigma} \le -\frac{1}{B}, \\ f_{j:n}^{-1}\left(1 + B\frac{x-\mu}{\sigma}\right), & \text{if } -\frac{1}{B} \le \frac{x-\mu}{\sigma} < \frac{f_{j:n}(\alpha_*)-1}{B}, \\ 1, & \text{if } \frac{x-\mu}{\sigma} \ge \frac{f_{j:n}(\alpha_*)-1}{B}. \end{cases} \quad (4.12)$$

For $j = n$, (4.2) is the best bound, attainable by (4.3).

Distribution function (4.12) has an absolutely continuous component (inverse of a polynomial of degree $n-1$), and a jump of height $1-\alpha_*$ at the right end. With few exceptions, the distribution does not have an explicit formula. Balakrishnan [10] used a relationship between binomial and negative binomial distributions for reducing polynomial equation (4.7) of degree n to one of degree $j-1$. This makes possible writing explicit values of solutions α_* and bounds $B^0(j,n)$ for $j = 2, 3, n-2, n-1$. For $j = 2$ in particular $\alpha_* = (n-1)^{-2}$, and (4.10) amounts to

$$||\bar{f}_{2:n}||^2 = \frac{n^2(n-1)}{(2n-3)(2n-1)}\left[1 - \frac{n^{2n-3}(n-2)^{2n-1}}{(n-1)^{4(n-1)}}\right]. \quad (4.13)$$

In Chapter 7 we need more rough approximations of $E_F X_{j:n}$ in m_F units. Observe that functions $P^\nearrow f_{j:n} = \bar{f}_{j:n}$ for $2 \leq j \leq n-1$ and $P^\nearrow f_{n:n} = f_{n:n}$ vanish at 0, and so coincide with the projections onto $\mathcal{C}^+ \subset \mathcal{C}^\nearrow$. Therefore, we have

$$E_F X_{j:n}/m_F \leq ||\bar{f}_{j:n}||, \quad 2 \leq j \leq n-1, \quad (4.14)$$
$$E_F X_{n:n}/m_F \leq ||f_{n:n}|| = n/(2n-1)^{1/2}, \quad (4.15)$$

(cf. (4.10) and (4.2), respectively). The bounds are attained by

$$F(x) = \begin{cases} 0, & \text{if } \frac{x}{m} \leq 0, \\ \bar{f}_{j:n}^{-1}(||\bar{f}_{j:n}||\frac{x}{m}), & \text{if } 0 \leq \frac{x}{m} < \frac{f_{j:n}(\alpha_*)}{||\bar{f}_{j:n}||}, \\ 1, & \text{if } \frac{x}{m} \geq \frac{f_{j:n}(\alpha_*)}{||\bar{f}_{j:n}||}, \end{cases} \quad (4.16)$$

for $2 \leq j \leq n-1$, and

$$F(x) = \begin{cases} 0, & \text{if } \frac{x}{m} \leq 0, \\ [(2n-1)^{-1/2}\frac{x}{m}]^{1/(n-1)}, & \text{if } 0 \leq \frac{x}{m} < (2n-1)^{1/2}, \\ 1, & \text{if } \frac{x}{m} \geq (2n-1)^{1/2} \end{cases} \quad (4.17)$$

for the sample maximum. Both (4.16) and (4.17) are life distributions. Density function $f_{1:n}$ is decreasing and so projection $P^\nearrow f_{1:n} = 1 \notin \mathcal{C}^+$, but the constant can be approximated in $L^2([0,1), dx)$ by functions

$$h_k(x) = (1 - k^{-1})^{-1/2} \mathbf{1}_{[1/k, 1)}(x).$$

This implies that the trivial bound for the sample minimum

$$\frac{E_F X_{1:n}}{m_F} \leq ||P^\nearrow f_{1:n}|| = 1 \quad (4.18)$$

is approximated by the sequence of two-point life distributions

$$P(X = 0) = k^{-1} = 1 - P(X = (1-k^{-1})^{-1/2} m_F), \quad k \to \infty. \quad (4.19)$$

The problem of determining bounds analogous to (4.11) for symmetric parent distributions can be solved by use of the trick of folding the functional about $1/2$. Case $j = n$ is the only one for which

$$s_{j:n}(x) = f_{j:n}(x) - f_{j:n}(1-x) = f_{j:n}(x) - f_{n+1-j:n}(x) \tag{4.20}$$

is nondecreasing. Explicit bound

$$\begin{aligned} E_F X_{n:n} - \mu &= n \int_{1/2}^{1} [F^{-1}(x) - \mu][x^{n-1} - (1-x)^{n-1}]\, dx \\ &\leq n \left[\int_{1/2}^{1} [x^{n-1} - (1-x)^{n-1}]^2 \, dx \right]^{1/2} \sigma/\sqrt{2} \\ &= \left\{ \frac{n^2}{2(2n-1)} \left[1 - \binom{2n-2}{n-1}^{-1} \right] \right\}^{1/2} \sigma \\ &= B^s(n,n)\sigma \sim n^{1/2}/2 \end{aligned} \tag{4.21}$$

(cf. (4.2)) was obtained by Moriguti [57] through use of the Schwarz inequality. The extreme distributions for which the equality holds have quantile functions equal to $x^{n-1} - (1-x)^{n-1}$ up to affine transformations.

Bounds for general order statistics of symmetric populations are derived by projecting differences $s_{j:n}$ onto the family of nondecreasing functions in $L^2([1/2,1), 2dx)$. For this purpose we first analyze variability of the differences, using the auxiliary results of Gajek and Rychlik [33, Lemma 3, p. 167].

Lemma 12 *Consider the function defined in (4.20) for $x \in [1/2, 1)$ and $(n+1)/2 < j < n$. This is nonnegative, and equal to 0 at $1/2$ and 1, and increasing-decreasing. For $j \leq n-2$ function (4.20) is concave-convex if $j \leq [n + (3n-5)^{1/2}]/2$, and convex-concave-convex otherwise. Also, $s_{n-1:n}$ is concave for $n \leq 7$, and convex-concave for $n \geq 8$.*

Now we merely apply the first statement of the lemma. Further properties are needed in Section 4.4. Observe first that, by symmetry, $s_{j:n} \leq 0$ for $j \leq (n+1)/2$, and so

$$E_F X_{j:n} - \mu = \int_{1/2}^{1} [F^{-1}(x) - \mu] s_{j:n}(x)\, dx \leq 0, \tag{4.22}$$

since $F^{-1} - \mu_F \geq 0$ on $[1/2, 1)$. If $j = (n+1)/2$, then $s_{j:n} = 0$, and (4.22) becomes the equality for any symmetric F. For $(n+1)/2 < j \leq n-1$, we repeat arguments leading to (4.8), and obtain

$$P^s f_{j:n}(x) = s_{j:n}(\min\{x, \alpha_*\}), \quad 1/2 \leq x < 1, \tag{4.23}$$

for $\alpha_* \in [1/2, 1)$ defined by equation

$$(1-\alpha)[f_{j:n}(\alpha) - f_{n+1-j:n}(\alpha)] = F_{n+1-j:n}(\alpha) - F_{j:n}(\alpha) \tag{4.24}$$

(cf. (4.7)). Using (4.9), we calculate

$$\begin{aligned}\|P^s f_{j:n}\|^2 &= n\frac{\binom{2j-2}{j-1}\binom{2n-2j}{n-j}}{\binom{2n-1}{n}}[F_{2j-1:2n-1}(\alpha_*) + F_{2n-2j+1:2n-1}(\alpha_*)\\&\quad - F_{2j-1:2n-1}\left(\frac{1}{2}\right) - F_{2n-2j+1:2n-1}\left(\frac{1}{2}\right)]\\&\quad - 2n\frac{\binom{n-1}{j-1}^2}{\binom{2n-1}{n}}\left[F_{n:2n-1}(\alpha_*) - \frac{1}{2}\right]\\&\quad + (1-\alpha_*)s_{j:n}^2(\alpha_*).\end{aligned} \quad (4.25)$$

Theorem 10 (symmetric distributions) *For $1 \leq j \leq (n+1)/2$, we have $\mathrm{E}_F X_{j:n} \leq \mu_F$ for all symmetric parent distributions of the sample. This becomes the equality if either F is the Dirac measure at μ_F or $j = (n+1)/2$.*

For $(n+1)/2 < j \leq n-1$, we have

$$\frac{\mathrm{E}_F X_{j:n} - \mu_F}{\sigma_F} \leq B = B^s(j,n) = \frac{\|P^s f_{j:n}\|}{\sqrt{2}}, \quad (4.26)$$

defined in (4.24) and (4.25), with the equality for

$$F(x) = \begin{cases} 0, & \text{if } \frac{x-\mu}{\sigma} < -\frac{s_{j:n}(\alpha_*)}{2B}, \\ s_{j:n}^{-1}(2B\frac{x-\mu}{\sigma}), & \text{if } -\frac{s_{j:n}(\alpha_*)}{2B} \leq \frac{x-\mu}{\sigma} < \frac{s_{j:n}(\alpha_*)}{2B}, \\ 1, & \text{if } \frac{x-\mu}{\sigma} \geq \frac{s_{j:n}(\alpha_*)}{2B}, \end{cases} \quad (4.27)$$

(see (4.23) and (4.25)).

For $j = n$ we have (4.21) which becomes the equality for (4.27) with $\alpha_* = 1$ and $s_{n:n}(1) = n$.

Distribution (4.27) has a smooth component with two atoms of measure $1 - \alpha_*$ ($= 0$ for $j = n$) at the endpoints of support.

4.2 Life Distributions with Decreasing Density and Failure Rate

Unlike the problems studied above, we present here bounds in terms of the square root of the second raw moment. In fact, except for the scale unit we also have a location parameter in the model. This is the population minimal value which for the life distributions amounts to 0. The problem is to evaluate the expected lifetime of $(n+1-j)$-out-of-n systems of independent elements whose identical distribution functions satisfy $F \succeq_c W$ for some fixed W. The solution is based on determining the projection $P_{\succeq_c W}^+ h$

4.2 Life Distributions with Decreasing Density and Failure Rate

of $h = f_{j:n} W$ onto the convex cone

$$\mathcal{C}^+_{\succeq_c W} = \{ g \in L^2([a_W, d_W), w(x)dx) : g - \text{nondecreasing, convex,} \\ g(a_W) = 0 \}. \quad (4.28)$$

In the sequel we frequently abstract from the specific form of h and assume the following

$$h(x), w(x) > 0 \quad \text{for } a < x < d, \quad h(a) = 0, \quad (4.29)$$

$$\int_a^d h(x) w(x) \, dx = \int_a^d w(x) \, dx = 1, \quad (4.30)$$

$$\int_a^d x^2 w(x) \, dx < \infty, \quad (4.31)$$

$$h(x) - \text{bounded}, \ h'' \text{ exists}, \quad (4.32)$$

$$\exists\, a \leq b < c \leq d \quad \begin{array}{l} \forall\, a < x < b \quad h'(x), h''(x) > 0, \\ \forall\, b < x < c \quad h''(x) < 0 < h'(x), \\ \forall\, c < x < d \quad h'(x) < 0. \end{array} \quad (4.33)$$

However, it is worth pointing out that for the representation $h = f_{j:n} W$, (4.29) through (4.33) are actually conditions on W, and some of them are naturally satisfied. For $j \neq 1$, we have (4.29) and (4.30), and boundedness and monotonicity assumptions (with $c = d$ for $j = n$). An equivalent formulation of (4.31) is finiteness of m_W^2, and existence of h'' requires differentiability of density w on (a_W, d_W). Crucial assumptions are ones describing regions of convexity and concavity of h. These were chosen to cover the important cases of uniform and exponential weights, without pretending to develop a general theory. One can easily check that the assumptions are satisfied for $2 \leq j \leq n-1$ and both $W = U, V$ (with $a = b$ in case $j = 2$), and for $j = n$ and $W = V$ with $c = d$. We point out that for $j = 1$ the trivial bound

$$E_F X_{1:n} \leq \mu_F \leq m_F$$

is valid for arbitrary F, being attainable for degenerate distributions. In the case $j = n$ and $W = U$, we have

$$P^+_{\succeq_c U} f_{n:n} = f_{n:n} \in \mathcal{C}^+_{\succeq_c U},$$

and so the bound for $F \succeq_c U$ is identical with that for general F.

We now present the solution to the projection problem. Lemma 13 describes geometrical properties of the projection. This allows us to restrict ourselves to continuous functions that are identical with h on the left and have linear extensions on the right.

4. Order Statistics of Independent Samples

Lemma 13 *For fixed h satisfying (4.29) through (4.33), set*

$$h_{\alpha\beta}(x) = \begin{cases} h(x), & \text{if } a \le x \le \beta, \\ h(\beta) + \alpha(x-\beta), & \text{if } \beta \le x < d. \end{cases} \quad (4.34)$$

For every $g \in \mathcal{C}^+_{\succeq_c W}$ there exists $h_{\alpha\beta} \in \mathcal{C}^+_{\succeq_c W}$ such that

$$\|h_{\alpha\beta} - h\| \le \|g - h\|.$$

PROOF. Take an arbitrary nonzero $g \in \mathcal{C}^+_{\succeq_c W}$. Since $g(a) = h(a) = 0$, it makes sense to define

$$\gamma = \sup\{x \in [a,b] : g(x) = h(x)\}.$$

We now analyze mutual relations between g and h in $[b,d)$, where h is first concave increasing and then possibly decreasing.

Suppose first that $g > h$ in a right neighborhood of b. Then these cases are possible:

(i) $g > h$ on the whole (b,d),

(ii) g and h are merely tangent in a single possibly degenerate subinterval of (b,c), or

(iii) g crosses h at a point $\delta \in (b,c)$, say, and either $g < h$ holds on (δ, d) or g and h cross each other at a single point $\Delta \in (\delta, d)$.

In each case we construct modifications g_i, $i = 1, 2, 3$, of the original g that have form (4.34) and are less distant from h than g.

(i) We have $g(x) > h(x)$ for all $x \in (\gamma, d)$. Define

$$\tilde{h}(x) = \begin{cases} h(b) + h'(b)(x-b), & \text{if } a \le x \le b, \\ h(x), & \text{if } b \le x \le c, \\ h(c), & \text{if } c \le x < d. \end{cases}$$

We easily see that $\tilde{h}$ is nondecreasing, concave, and $\tilde{h} < g$. Therefore the regions above the graph of g and beneath the graph of $\tilde{h}$ are disjoint planar sets, and there exists a straight line $l_1(x)$ separating the sets from each other. This line must have a common point, say $\beta \in [\gamma, b]$, with h. Then $g_1 = \max\{h, l_1\}$ has the desired form. Moreover, $\|g_1 - h\| \le \|g - h\|$, because $h(x) = g_1(x)$ for $x \in [a, \beta]$ and $h(x) \le g_1(x) \le g(x)$ for $x \in [\beta, d)$.

(ii) Functions h and $g \ge h$ are concave and convex, respectively, in (b,c), and have a tangent point there. Accordingly, there exists a line l_2 separating the curves in the interval. Note that l_2 crosses h at some $\beta \in [\gamma, b)$, because it runs beneath g and over the concave part of h, and $h(\gamma) = g(\gamma)$. For $g_2 = \max\{h, l_2\}$ we easily verify the conclusions of the previous case.

4.2 Life Distributions with Decreasing Density and Failure Rate

(*iii*) Extend the definition of Δ as follows

$$\Delta = \sup\{x \geq \delta : g(x) \leq h(x)\}.$$

In the former subcase $\Delta = d$ which is possible only if $d < \infty$. Note that

$$g(\Delta) = \lim_{x \nearrow \Delta} g(x),$$
$$h(\Delta) = \lim_{x \nearrow \Delta} h(x),$$

are well defined and finite. Define

$$l_{\delta\Delta}(x) = g(\delta) + \frac{g(\Delta) - g(\delta)}{\Delta - \delta}(x - \delta). \tag{4.35}$$

Then we have

$$g(x) \leq l_{\delta\Delta}(x) \leq h(x), \quad \text{if} \quad \delta \leq x \leq \Delta,$$
$$h(x) \leq l_{\delta\Delta}(x) \leq g(x), \quad \text{if} \quad \Delta \leq x < d,$$

and the latter holds in a left neighborhood of δ, as well. By convexity of g, we have

$$l_{\delta\Delta}(x) \leq g(x), \quad a \leq x \leq \delta.$$

This relation, combined with $g(\gamma) = h(\gamma)$, implies the existence of $\beta \in [\gamma, \delta]$ such that $l_{\delta\Delta}(\beta) = h(\beta)$ and $l'_{\delta\Delta} \geq h'(\beta)$. Therefore

$$g_3(x) = \begin{cases} h(x), & \text{if } a \leq x \leq \beta, \\ l_{\delta\Delta}(x), & \text{if } \beta \leq x < d, \end{cases}$$

is the desired convex nondecreasing modification of g.

If $g < h$ on some right neighborhood of b (and so of γ), then there is a single point $\Delta > b$ at most at which g and h cross each other. If there is no such Δ, we set $\Delta = d$, being finite (cf. Case (*iii*)). For each $y \in [a, b)$, define the linear function tangent to h at y as

$$l_y(x) = h(y) + h'(y)(x - y). \tag{4.36}$$

Notice that the slopes of (4.36) increase, and $y_1 < y_2$ implies $l_{y_1} < l_{y_2}$ on $[y_2, d)$. Function $y \mapsto l_y(\delta)$, $y \in [a, b]$, is strictly increasing and continuous, and satisfies

$$l_b(\Delta) \geq h(\Delta) \geq g(\Delta).$$

Moreover, relation $l_\gamma(\Delta) > h(\Delta)$ implies

$$l'_\gamma = h'(\gamma) > \frac{h(\Delta) - h(\gamma)}{\Delta - \gamma} \geq \frac{g(\Delta) - g(\gamma)}{\Delta - \gamma} \geq g'(\gamma+). \tag{4.37}$$

If $\gamma > a$, then, by (4.37), we have $h'(\gamma) < g'(\gamma-)$, and $h < g$ on a left neighborhood of γ. Summing up, under condition $g(\gamma+) < h(\gamma+)$, three cases are possible:

(iv) $l_\gamma(\Delta) \leq h(\Delta)$,

(v) $l_\gamma(\Delta) > h(\Delta)$ with $\gamma = a$, and

(vi) $l_\gamma(\Delta) > h(\Delta)$ with $g(\gamma-) > h(\gamma-)$,

which are considered consecutively in the remainder of the proof.

(iv) Relations $l_\gamma(\Delta) \leq h(\Delta) \leq l_b(\Delta)$ imply existence of a $\delta \in [\gamma, b]$ for which $l_\delta(\Delta) = h(\Delta)$ holds. Then

$$g(x) \leq l_\delta(x) \leq h(x), \qquad x \in [\delta, \Delta].$$

If $(\Delta, d) \neq \emptyset$, then the reversed inequalities hold there. Therefore the convex nondecreasing function

$$g_4(x) = \max\{h(x), l_\delta(x)\} = h(x)\mathbf{1}_{[a,\delta]}(x) + l_\delta(x)\mathbf{1}_{(\delta,d)}(x)$$

is a better approximation of h than g.

(v) The linear function $g_5(x) = l_{\gamma\Delta}(x)$ secant to g at $\gamma = a$ and Δ (cf. (4.35)) runs between g and h in the whole domain $[a, d)$.

(vi) In this case there exists $\delta \in [a, \gamma)$ such that $g(\delta) = h(\delta)$, and $g(x) > h(x)$ for $x \in (\delta, \gamma)$. Apply (4.35) again for defining the straight line $l_{\delta\Delta}$. We have

$$g(x) \leq l_{\delta\Delta}(x) \leq h(x), \qquad x \in [\gamma, \delta],$$

and

$$g(x) \geq l_{\delta\Delta}(x) \geq h(x), \qquad x \in [\Delta, d),$$

under condition $\Delta < d$ (cf. Case (iv)). For $x < \gamma$, we proceed as in Case (iii). We have

$$h(\gamma-) < l_{\delta\Delta}(\gamma-) \leq g(\gamma-),$$
$$l_{\delta\Delta}(\delta) \leq \quad g(\delta) \quad = h(\delta),$$

which imply that $l_{\delta\Delta}(\delta_0) = h(\delta_0)$ for a unique $\delta_0 \in [\delta, \gamma)$, and

$$h(x) < l_{\delta\Delta}(x) \leq g(x), \qquad x \in [\delta_0, \gamma).$$

It follows that $h'(\delta_0) < l'_{\delta\Delta}$, and

$$g_6(x) = h(x)\mathbf{1}_{[a,\delta_0]}(x) + l_{\delta\Delta}(x)\mathbf{1}_{(\delta_0,d)}(x)$$

is an element of $\mathcal{C}^+_{\succeq_c W}$ that lies closer to h than the original g. This ends the proof. ∎

Observe that $h_{\alpha\beta} \in \mathcal{C}^+_{\succeq_c W}$ requires $a \leq \beta \leq b$, and either $\alpha \geq h'(\beta) > 0$ for $\beta > a$ or $\alpha \geq 0$ for $\beta = a$. If $a = b$ in particular, then the projection is a nondecreasing linear function

$$P^+_{\succeq_c W} h(x) = \alpha_*(x - a)$$

4.2 Life Distributions with Decreasing Density and Failure Rate

whose slope α_* can be easily determined. Note that for fixed $\beta \in [a,b]$,

$$D(\alpha,\beta) = ||h_{\alpha\beta} - h||^2 = \int_\beta^d [\alpha(x-\beta) + h(\beta) - h(x)]^2 w(x)\, dx \qquad (4.38)$$

is a convex quadratic function of α, and under restriction $h_{\alpha\beta} \in \mathcal{C}^+_{\succeq_c W}$ is minimized at

$$\hat{\alpha}(a) = \max\{0, \alpha_*(a)\}, \qquad (4.39)$$
$$\hat{\alpha}(\beta) = \max\{h'(\beta), \alpha_*(\beta)\}, \quad \beta > a, \qquad (4.40)$$

with

$$\alpha_*(\beta) = \frac{\int_\beta^d (x-\beta)[h(x) - h(\beta)] w(x)\, dx}{\int_\beta^d (x-\beta)^2 w(x)\, dx}, \qquad (4.41)$$

being the optimal slope without restrictions. Consecutive reasoning steps consist in eliminating $\beta > a$ for which $h'(\beta) > \alpha_*(\beta)$, and analyzing $D(\alpha_*(\beta), \beta)$ with continuous derivative

$$\frac{dD(\alpha_*(\beta), \beta)}{d\beta} = 2[\alpha_*(\beta) - h'(\beta)] \int_\beta^d [h(x) - h_{\alpha_*(\beta)\beta}(x)] w(x)\, dx$$
$$= 2K(\beta) L(\beta), \qquad (4.42)$$

say. We minimize $D(\alpha_*(\beta), \beta)$ by determining the set

$$\mathcal{K} = \{\beta \in (a,b) : K(\beta) \geq 0\},$$

and analyzing sign changes of L in $\mathcal{K}$. If $\mathcal{K} = (a, \kappa]$ for some $\kappa \in (a,b]$ (which actually holds in special cases $W = U, V$), then $L(\beta)$ is positive, negative, or negative-positive, with zero at some $\lambda \in \mathcal{K}$, and so optimal $\beta_* = a, \kappa$, and λ in the respective cases. Specific forms of

$$h(x) = f_{j:n}(x) = n B_{j-1, n-1}(x),$$

where

$$B_{k,m}(x) = \binom{m}{k} x^k (1-x)^{m-k}, \quad 0 \leq k \leq m < \infty,$$

are the *Bernstein polynomials* of degree m, enable us to determine final forms of $P^+_{\succeq_c U} f_{j:n}$. Differentiating and integrating the Bernstein polynomials we obtain linear combinations of these polynomials of smaller and greater degree, respectively. These operations were applied in defining factors of (4.42) which coincide with combinations of Bernstein polynomials of degree $n+1$ up to positive functional multiplicators. The coefficients of the combinations depend on j and n. The basic tool in analyzing the sign changes of the combinations is the following well-known variation diminishing property of the Bernstein polynomials proved in Schoenberg [93, p. 252]).

4. Order Statistics of Independent Samples

Lemma 14 *The number of zeros of a given nonzero combination of Bernstein polynomials*

$$B(x) = \sum_{k=0}^{m} a_k B_{k,m}(x), \quad x \in (0,1), \tag{4.43}$$

does not exceed the number of sign changes of the sequence $a_0, \ldots, a_m$. *The first and last signs of* (4.43) *are identical with the signs of the first and last nonzero elements of* $a_0, \ldots, a_m$, *respectively.*

This is equivalent to saying that $B_{k,m}$, $0 \leq k \leq m$, for fixed m form a Chebyshev system on $[0,1]$ (see Karlin and Studden [45, Theorem 1.4.1]). In addition, differentiation and integration operators defined in (4.41) and (4.42), acting on compositions

$$h(x) = f_{j:n} V(x) = n B_{j-1, n-1} V(x),$$

do not lead beyond the polynomials of degree n in $V(x)$ up to positive factors and can be further rewritten as combinations of $B_{k,n} V(x)$. Therefore Lemma 14 is useful in analyzing variability of (4.42) for $h(x) = f_{j:n} V(x)$ as well.

Theorem 11 (decreasing density) *If* $2 \leq j \leq \min\{2(n+1)/3, n-1\}$, *then*

$$\frac{\mathrm{E}_F X_{j:n}}{m_F} \leq \sqrt{3} \frac{j}{n+1}, \tag{4.44}$$

which is the equality if F is the uniform distribution on $[0, \sqrt{3} m_F]$.
If $2(n+1)/3 < j \leq n-1$, *then*

$$\frac{\mathrm{E}_F X_{j:n}}{m_F} \leq B = B^+_{\succeq_c U}(j, n) = \|(f_{j:n})_{\alpha_* \beta_*}\|, \tag{4.45}$$

(cf. (4.34)*) for*

$$\|(f_{j:n})_{\alpha_* \beta_*}\|^2 = n \frac{\binom{2j-2}{j-1}\binom{2n-2j}{n-j}}{\binom{2n-1}{n}} F_{2j-1:2n-1}(\beta_*) + (1-\beta_*) f_{j:n}^2(\beta_*)$$
$$+ (1-\beta_*)^2 \alpha_* f_{j:n}(\beta_*) + (1-\beta_*)^3 \alpha_*^2/3, \tag{4.46}$$

$$\alpha_* = \alpha_*(\beta_*) = \frac{3}{(1-\beta_*)^3} \left\{ \frac{j}{n+1}[1 - F_{j+1:n+1}(\beta_*)] \right.$$
$$\left. - \beta_*[1 - F_{j:n}(\beta_*)] \right\} - \frac{3}{2(1-\beta_*)} f_{j:n}(\beta_*), \tag{4.47}$$

and $\beta_* = \beta_*(j,n)$ *being the smaller of the smallest positive zeros of polynomials*

4.2 Life Distributions with Decreasing Density and Failure Rate

$$K_U(x) = 3\sum_{k=1}^{j}(j+1-k)f_{k:n+2}(x) - \frac{(n-j+3)!}{(n-j)!}f_{j-1:n+2}(x)$$
$$+ (n-j-\frac{3}{2})\frac{(n-j+2)!}{(n-j)!}f_{j:n+2}(x), \tag{4.48}$$

$$L_U(x) = \sum_{k=1}^{j-1}(n+1-\frac{3j-k}{2})f_{k:n+2}(x)$$
$$- \frac{(n-j)(n-j-1)}{4}f_{j:n+2}(x). \tag{4.49}$$

Bound (4.45) is attained by

$$F(x) = \begin{cases} 0, & \text{if } \frac{x}{m} \le 0, \\ f_{j:n}^{-1}\left(B\frac{x}{m}\right), & \text{if } 0 \le \frac{x}{m} \le \frac{f_{j:n}(\beta_*)}{B}, \\ \beta_* + \frac{B\frac{x}{m} - f_{j:n}(\beta_*)}{\alpha_*}, & \text{if } \frac{f_{j:n}(\beta_*)}{B} \le \frac{x}{m} \le \frac{f_{j:n}(\beta_*)+\alpha_*(1-\beta_*)}{B}, \\ 1, & \text{if } \frac{x}{m} \ge \frac{f_{j:n}(\beta_*)+\alpha_*(1-\beta_*)}{B}. \end{cases} \tag{4.50}$$

Finally, for $j = n$ we have (4.15) which becomes the equality for (4.17).

Theorem 12 (decreasing failure rate) *Set*

$$\mu_{V_{j:n}} = E_V X_{j:n} = \sum_{k=1}^{j}\frac{1}{n+1-k}, \quad 1 \le j \le n. \tag{4.51}$$

If $\mu_{V_{j:n}} \le 2$, then

$$\frac{E_F X_{i:n}}{m_F} \le \frac{\mu_{V_{j:n}}}{\sqrt{2}}, \tag{4.52}$$

which becomes the equality for the exponential distribution with the scale parameter $m_F/\sqrt{2}$.

Otherwise

$$\frac{E_F X_{j:n}}{m_F} \le B = B^+_{\succeq_c V}(j,n) = \|(f_{j:n}V)_{\alpha_* V^{-1}(\gamma_*)}\|, \tag{4.53}$$

for

$$\|(f_{j:n}V)_{\alpha_* V^{-1}(\gamma_*)}\|^2 = n\frac{\binom{2j-2}{j-1}\binom{2n-2j}{n-j}}{\binom{2n-1}{n}}F_{2j-1:2n-1}(\gamma_*)$$
$$+ (1-\gamma_*)[2\alpha_*^2 + 2\alpha_* f_{j:n}(\gamma_*) + f_{j:n}^2(\gamma_*)], \tag{4.54}$$

$$\alpha_* = \alpha_*(V^{-1}(\gamma_*)) = \sum_{k=1}^{j}\frac{\mu_{V_{j+1-k:n+1-k}}}{2n(1-\gamma_*)}f_{k:n+1}(\gamma_*)$$
$$- \frac{1}{2}f_{j:n+1}(\gamma_*), \tag{4.55}$$

and γ_* being the minimum of the smallest positive zeros of

$$K_V(x) = \sum_{k=1}^{j} \mu_{V_{j+1-k:n+1-k}} f_{k:n+1}(x)$$
$$- 2\frac{(n-j+2)!}{(n-j)!} f_{j-1:n+1}(x)$$
$$+ 2(n-j-\frac{1}{2})(n-j+1) f_{j:n+1}(x), \qquad (4.56)$$

$$L_V(x) = \sum_{k=1}^{j} [2 - \mu_{V_{j+1-k:n+1-k}}] f_{k:n+1}(x)$$
$$- (n+1-j) f_{j:n+1}(x). \qquad (4.57)$$

Bound (4.53) is attained by

$$F(x) = \begin{cases} 0, & \text{if } \frac{x}{m} \leq 0, \\ f_{j:n}^{-1}\left(B\frac{x}{m}\right), & \text{if } 0 \leq \frac{x}{m} \leq \frac{f_{j:n}(\gamma_*)}{B}, \\ 1 - (1-\gamma_*)\exp\left(-\frac{B\frac{x}{m} - f_{j:n}(\gamma_*)}{\alpha_*}\right), & \text{if } \frac{x}{m} \geq \frac{f_{j:n}(\gamma_*)}{B}. \end{cases} \qquad (4.58)$$

Formulae (4.45) through (4.47), and (4.53) through (4.55) are concluded from (4.34) and (4.41). Moreover, (4.48) and (4.49) as well as (4.56) and (4.57) differ from K and L defined in (4.42) by positive functional factors common for all j, and original variables are replaced by $V(x)$ in the latter case. All bounds are achieved by absolutely continuous distributions. The left-hand parts of (4.50) and (4.58) are the inverses of polynomials equal to $f_{j:n}$ up to scale factors, and cannot be written explicitly. They have uniform and exponential right tails, respectively. Bound (4.15) was derived directly from the Schwarz inequality. Using the following integral approximations of harmonic series

$$\ln \frac{n+1}{n+1-j} < \mu_{V_{j:n}} < \ln \frac{n+1/2}{n+1/2-j},$$

we deduce that condition $\mu_{V_{j:n}} \leq 2$ providing bound (4.52) is true for $j \leq (1-e^{-2})(n+1/2)$ and false for $j \geq (1-e^{-2})(n+1)$. The difference between both estimates is $(1-e^{-2})/2 \approx 0.43233$ which implies that for given n the condition has to be directly checked for one j at most. Bounds of Theorem 12 are tighter than those of Theorem 11, because a narrower class of distributions was treated there. This is confirmed by numerical comparisons presented in Gajek and Rychlik [33, Table I, p. 162].

4.3 Distributions with Monotone Density and Failure Rate on the Average

The crucial points in calculating mean-variance bounds for $E_F X_{j:n}$ for F with decreasing density and failure rate on the average lies in deriving projections $P^0_{\succeq_* U}(f_{j:n}-1)$ and $P^0_{\succeq_* V}(f_{j:n}V-1)$, respectively, onto convex cones of nondecreasing starshaped functions that are orthogonal to constants. In fact, we consider a more general problem of projecting $h = f_{j:n}W$ satisfying (4.29) to (4.33) onto $\mathcal{C}^{\nearrow}_{\succeq_* W}$ which gives

$$P^0_{\succeq_* W}(h-1) = P^{\nearrow}_{\succeq_* W} h - 1.$$

Note first that
$$P^{\nearrow}_{\succeq_* W} h(a) \in [h(a), h(c)).$$

Functions starting from a higher level uniformly majorize h, and contradict the statement of Lemma 2. Ones with $g(a) < h(a)$ are eliminated by

$$g_{h(a)} = \max\{g, h(a)\} \in \mathcal{C}^{\nearrow}_{\succeq_* W}.$$

Therefore in Lemma 15, describing the shape of projection functions, we confine ourselves to functions starting from a range point of h.

Lemma 15 *Define functions*

$$h_{\beta\alpha}(x) = \begin{cases} h(\beta), & \text{if } a \leq x \leq \beta, \\ h(x), & \text{if } \beta \leq x \leq \alpha, \\ \frac{h(\alpha)-h(\beta)}{\alpha-a}(x-a) + h(\beta), & \text{if } \alpha \leq x < d, \end{cases} \quad (4.59)$$

for some $\alpha \neq a \leq \beta \leq \alpha < c$. *For every* $a \leq \beta < c$ *and* $g \in \mathcal{C}^{\nearrow}_{\succeq_* W}$ *satisfying* $g(a) = h(\beta)$ *there exists* $a \neq \alpha \in [\beta, c)$ *such that* $h_{\beta\alpha} \in \mathcal{C}^{\nearrow}_{\succeq_* W}$ *and*

$$\|h_{\beta\alpha} - h\| \leq \|g - h\|. \quad (4.60)$$

Functions (4.59) are obviously nondecreasing, and constant functions

$$h_{\beta\beta}(x) \equiv h(\beta), \quad a < \beta < c,$$

are also possible here. However, starshapedness is not apparent here. For fixed $\beta \in [a,c)$, let

$$s_\beta(\alpha) = \frac{h(\alpha) - h(\beta)}{\alpha - a}, \quad a \neq \alpha \in [\beta, c), \quad (4.61)$$

denote the slopes of lines

$$l_{BA}(x) = s_\beta(\alpha)(x - a) + h(\beta) \quad (4.62)$$

passing through $B = (a, h(\beta))$ and points $A = (\alpha, h(\alpha))$, $\beta \leq \alpha < c$, of the graph of h (cf. the last line of (4.59)). Write

$$s_a(a) = h'(a) \geq 0$$

for completeness. We first prove the following.

Lemma 16 *For every $\beta \in [a, c)$ there exists $\bar{\alpha}(\beta) \in \mathcal{B}(\beta) = [\max\{\beta, b\}, c)$ such that (4.61) increases on $(\beta, \bar{\alpha}(\beta))$ and decreases on $(\bar{\alpha}(\beta), c)$.*

In consequence, (4.59) is actually starshaped if $\alpha \in [\beta, \bar{\alpha}(\beta)]$. Observe that $\bar{\alpha}(\beta)$ is the point where $l_{B\bar{A}(\beta)}$ is tangent to h and therefore can be determined by

$$h(\alpha) - h(\beta) = h'(\alpha)(\alpha - a). \tag{4.63}$$

PROOF OF LEMMA 16. The statement is trivial for $\beta = a = b$ with $\bar{\alpha}(\beta) = a$. Indeed, by concavity of h on (a, c), for any $a < \alpha_1 < \alpha_2 < c$ point $A_1 = (\alpha_1, h(\alpha_1))$ lies above l_{BA_2} (cf. (4.62)) and so $s_a(\alpha_1) > s_a(\alpha_2)$.

We now claim that s_β is increasing-decreasing on $\mathcal{B}(\beta)$ if either $\beta > a = b$ or $\beta \geq a \neq b$. The proof consists in checking the falsity of the contradiction: for every fixed $\alpha \in \mathcal{B}(\beta)$ there exist $\alpha_1 < \alpha < \alpha_2$, $\alpha_i \in \mathcal{B}(\beta)$, such that $s_\beta(\alpha_i) \geq s_\beta(\alpha)$, $i = 1, 2$. If l_{BA} is tangent to h at α, it runs above all graph points A', $\alpha \neq \alpha' \in \mathcal{B}(\beta)$. Accordingly, $s_\beta(\alpha') < s_\beta(\alpha)$. If l_{BA} is a secant line to the curve, then these cross each other once more in $\mathcal{B}(\beta)$ at most. If $h - l_{BA}$ changes its sign from $-$ to $+$ at α, then for all $\mathcal{B}(\beta) \ni \alpha_1 < \alpha$ we have

$$h(\alpha_1) < l_{BA}(\alpha_1),$$
$$s_\beta(\alpha_1) < s_\beta(\alpha).$$

If the sign changes from $+$ to $-$, then the analogous relations hold for $\mathcal{B}(\beta) \ni \alpha_2 > \alpha$. Therefore our claim is actually true.

By continuity of s_β, it remains to show that (4.61) is increasing in (β, b) when $a \leq \beta < b$. Note that

$$h_{h(\beta)}(x) = \max\{h(\beta), h(x)\}$$

is convex in (a, b) and strictly convex in (β, b). For every $\beta \leq \alpha_1 < \alpha_2 \leq b$ point A_1 lies below l_{BA_2}, and hence $s_\beta(\alpha_1) < s_\beta(\alpha_2)$. This completes the proof. ∎

PROOF OF LEMMA 15. If $g(\bar{\alpha}(\beta)) \geq h(\bar{\alpha}(\beta))$, then we can take $h_{\beta\bar{\alpha}(\beta)}$. Since $h(x) < h(\beta)$ on $[a, \beta)$, we have $h < h_{\beta\bar{\alpha}(\beta)} \leq g$ there, because constant $h(\beta)$ is the smallest nondecreasing function starting from level $h(\beta)$. Also, $h_{\beta\bar{\alpha}(\beta)} = h$ is the optimal approximation of h on $[\beta, \bar{\alpha}(\beta)]$. Finally, for $x > \bar{\alpha}(\beta)$, we have

$$h(x) \leq \frac{h(\bar{\alpha}(\beta)) - h(\beta)}{\bar{\alpha}(\beta) - a}(x - a) + h(\beta) \leq g(x).$$

4.3 Distributions with Monotone Density and Failure Rate on the Average

The former inequality holds, because the middle term defines the line that is tangent to h at $\bar{\alpha}(\beta)$ and majorizes h. The latter is a consequence of the fact that the line is the smallest starshaped function in $[\bar{\alpha}(\beta), d)$ that passes through B and $\bar{A}(\beta) = (\bar{\alpha}(\beta), h(\bar{\alpha}(\beta)))$. The above arguments show that $h_{\beta\bar{\alpha}(\beta)}$ actually satisfies (4.60).

However, $h \neq h_{\beta\bar{\alpha}(\beta)} \geq h$ and, due to Lemma 2, the approximation can be further improved by a downward translation of $h_{\beta\bar{\alpha}(\beta)}$. We consider the case $g(\bar{\alpha}(\beta)) < h(\bar{\alpha}(\beta))$ now. Define

$$\delta = \inf\{x > \bar{\alpha}(\beta) : g(x) \geq h(x)\},$$

putting $\delta = d$ if $g < h$ in $(\bar{\alpha}(\beta), d)$. Take

$$l_{BD}(x) = \frac{h(\delta) - h(\beta)}{\delta - a}(x - a) + h(\beta).$$

We first prove that $l_{BD} < h$ in $(\bar{\alpha}(\beta), \delta)$. If $(c, \delta) \neq \emptyset$, then $l_{BD} - h$ is strictly increasing there, and $l_{BD}(\delta) - h(\delta) \leq 0$. For $x \in (\bar{\alpha}(\beta), \min\{c, \delta\})$ the inequality follows from the strict convexity of $l_{BD} - h$ in the interval, and its nonpositivity at the endpoints. Now we check that $l_{BD} - h$ changes the sign in $(a, \bar{\alpha}(\beta))$ once at most. Assume that α is the largest point of sign change in $(a, \bar{\alpha}(\beta))$. Evidently $l_{BD} - h$ is positive and negative on the left and right to α, respectively. If $\alpha \in [b, \bar{\alpha}(\beta))$, then convexity of $l_{BD} - h$ implies its positivity in (b, α). This is also positive in (a, b) by its concavity there, and nonnegativity at the endpoints. If $\alpha \in (a, b)$, it suffices to repeat the above argument with b replaced by α. Assume that unique α exists. We can write

$$l_{BD}(x) = \frac{h(\alpha) - h(\beta)}{\alpha - a}(x - a) + h(\beta).$$

Since $l_{BD} \geq h(\beta)$, we get $\alpha \in [\beta, \bar{\alpha}(\beta))$. Note that in the class of nondecreasing starshaped functions passing through $B = (a, h(\beta))$ and $D = (\delta, g(\delta))$, function l_{BD} is maximal in $[a, \delta]$ and minimal in $[\delta, d)$. Therefore

$$g(x) \leq l_{BD}(x) \leq h(x), \quad \text{if} \quad x \in [\alpha, \delta],$$
$$h(x) \leq l_{BD}(x) \leq g(x), \quad \text{if} \quad x \in [\delta, d).$$

Also, we have

$$g(x) \leq h(\beta) \leq g(x), \quad x \in [a, \beta].$$

Consequently, $h_{\beta\alpha}$ defined as $h(\beta)$, h, and l_{BD} in $[a, \beta]$, $[\beta, \alpha]$, and $[\alpha, d)$, respectively, lies closer to h than g. Since $\beta \leq \alpha < \bar{\alpha}(\beta)$, we see that $h_{\beta\alpha} \in \mathcal{C}^{\nearrow}_{\succeq_* W}$.

If $\beta = a$, it may happen that l_{BD} has no sign changes in $[a, \delta)$. Then $l_{BD}(a) = h(a) = 0$ and $l_{BD} < h$ in (a, δ), and $l_{BD} > h$ in (δ, d). The last relation is a consequence of concavity and ultimate decrease of h. If $l_{BD} = 0 \leq h$, referring to Lemma 2 we decrease the distance to h by adding a positive constant l_{BD}^+. If l_{BD} has a positive slope, we can take a line l_{BD}^- running through $(\delta, g(\delta))$ with a slightly smaller slope. Observe that l_{BD}^- lies closer to h than the original one in $[a, \delta]$ and $[\delta, d)$. Both modifications lead to linear functions that cross h once at some $\alpha < \bar{\alpha}(\beta)$. Therefore we are in position to apply the construction of the previous paragraph, which ends the proof. ∎

Below we determine the optimal parameters. For $a \leq \beta \leq c$ and $\beta \leq \alpha \leq c$, $\alpha \neq a$, set

$$K(\alpha, \beta) = \int_\alpha^d [h_{\alpha\beta}(x) - h(x)](x-a)w(x)dx, \qquad (4.64)$$

$$L(\alpha, \beta) = \int_a^d [h_{\alpha\beta}(x) - h(x)]w(x)\,dx. \qquad (4.65)$$

Then for $a < \beta \leq c$ write

$$\hat{K}(\beta) = K(\beta, \beta) = \int_\beta^d [h(\beta) - h(x)](x-a)w(x)\,dx, \qquad (4.66)$$

$$\hat{L}(\beta) = L(\beta, \beta) = \int_a^d [h(\beta) - h(x)]w(x)\,dx = h(\beta) - 1. \qquad (4.67)$$

Lemma 17 *Let $\hat{\beta}$ be the unique zero of (4.67) in (a, c). If*

$$\hat{K}(\hat{\beta}) = \int_{\hat{\beta}}^d [1 - h(x)](x-a)w(x)\,dx \geq 0, \qquad (4.68)$$

then $P^\nearrow_{\succeq_* W} h = h_{\hat{\beta}\hat{\beta}} = 1$.

Otherwise there exists a unique pair (β_*, α_*), $a \leq \beta_* < \hat{\beta}$, $\max\{\beta_*, b\} < \alpha_* < c$, determined by equations

$$K(\alpha, \beta) = 0, \qquad (4.69)$$
$$L(\alpha, \beta) = 0, \qquad (4.70)$$

such that $P^\nearrow_{\succeq_* W} h = h_{\alpha_* \beta_*}$, defined as in (4.59).

Since h is strictly increasing from $h(a) = 0$ to $h(c) = \sup h > 1$ (cf. (4.29) and (4.30)), $\hat{\beta}$ is actually well defined.

PROOF OF LEMMA 17. By Lemma 15, we should minimize

$$D(\beta, \alpha) = \int_a^d [h_{\beta\alpha}(x) - h(x)]^2 w(x)\,dx$$

$$= \int_a^\beta [h(\beta) - h(x)]^2 w(x)\,dx$$

$$+ \int_\alpha^d [l_{BA}(x) - h(x)]^2 w(x)\,dx \qquad (4.71)$$

4.3 Distributions with Monotone Density and Failure Rate on the Average

(cf. (4.61) and (4.62)) with respect to two parameters $\beta \in [a,c)$ and $a \neq \alpha \in [\max\{\beta,b\}, \bar{\alpha}(\beta)) \subset (a,c)$. Fixing β and differentiating (4.71) with respect to α, we obtain

$$\frac{\partial D(\beta,\alpha)}{\partial \alpha} = \frac{2}{\alpha - a}[h'(\alpha) - l'_{BA}] \int_\alpha^d [l_{BA}(x) - h(x)]w(x)\,dx$$

$$= \frac{2}{\alpha - a}[h'(\alpha) - l'_{BA}]K(\beta,\alpha) \qquad (4.72)$$

(cf (4.64)). If $\beta < \alpha < \bar{\alpha}(\beta)$, then $h - l_{BA}$ changes the sign from $-$ to $+$ at α and hence the expression in brackets is positive. This vanishes at $\bar{\alpha}(\beta)$, but $\bar{\alpha}(\beta)$ cannot be optimal, because $h_{\beta\bar{\alpha}(\beta)} \geq h$. Analyzing the sign of (4.72), it suffices to concentrate on (4.64). We have

$$\frac{\partial K(\beta,\alpha)}{\partial \alpha} = \frac{h'(\alpha) - l'_{BA}}{\alpha - a} \int_\alpha^d (x-a)^2 w(x)\,dx > 0,$$

which implies that (4.72) is the product of a positive function and increasing $K(\beta, \cdot)$. Since the integrand is positive for $\alpha = \bar{\alpha}(\beta)$, we have $K(\beta, \bar{\alpha}(\beta)) > 0$. If $\alpha = \beta > a$, then $l_{BA} = h(\beta)$. It follows that $\hat{K}(\beta) < 0$ as $\beta \searrow a$ and $\hat{K}(\beta) > 0$ for $\beta \nearrow c$.

We can summarize the behavior of (4.71) as follows. If β is small enough then (4.72) is negative for α close to β, and changes its sign at an $\alpha_*(\beta) \in (\beta, \bar{\alpha}(\beta))$ where (4.64) vanishes and the unique minimum of $D(\beta, \cdot)$ is attained. If $\beta \geq \tilde{\beta}$ satisfying $\hat{K}(\tilde{\beta}) = 0$ then (4.72) is positive for all $\alpha > \beta$. Then $D(\beta, \cdot)$ is minimized at $\alpha_*(\beta) = \beta$ which gives a constant approximation $h_{\beta\beta} = h(\beta)$. It remains to choose $\beta \in [a,c)$ such that $(\beta, \alpha_*(\beta))$ minimizes (4.71), where $\alpha_*(\beta) > \beta$ satisfies $K(\beta, \alpha_*(\beta)) = 0$ for $\beta < \tilde{\beta}$ and $\alpha_*(\beta) = \beta$ for $\beta \geq \tilde{\beta}$.

By Lemma 2, a necessary condition for that is

$$L(\beta, \alpha_*(\beta)) = \int_a^d [h_{\beta\alpha_*(\beta)}(x) - h(x)]w(x)\,dx = 0 \qquad (4.73)$$

(cf (4.65)). It is clear that (4.67) strictly increases from negative $\hat{L}(a)$ to positive $\hat{L}(c) = L(c, \alpha_*(c))$. We have $\hat{L}(\hat{\beta}) = 0$ if

$$h_{\hat{\beta}\hat{\beta}} = h(\hat{\beta}) = \int_a^d h(x)w(x)\,dx = 1.$$

We show that $L(\beta, \alpha_*(\beta))$ is also increasing, when $\alpha_*(\beta) > \beta$ is determined by (4.69). Consider

$$\frac{dL(\beta, \alpha_*(\beta))}{d\beta} = h'(\beta) \left[\int_a^\beta w(x)\, dx + \int_{\alpha_*(\beta)}^d w(x)\, dx \right.$$
$$\left. - \int_{\alpha_*(\beta)}^d \frac{x-a}{\alpha_*(\beta) - a} w(x)\, dx \right]$$
$$+ \frac{d\alpha_*(\beta)}{d\beta} [h'(\alpha_*(\beta)) - l'_{BA_*(\beta)}]$$
$$\times \int_{\alpha_*(\beta)}^d \frac{x-a}{\alpha_*(\beta) - a} w(x)\, dx. \qquad (4.74)$$

Plugging

$$\frac{d\alpha_*(\beta)}{d\beta} = \frac{\frac{\partial K(\beta, \alpha_*(\beta))}{\partial \beta}}{\frac{\partial K(\beta, \alpha_*(\beta))}{\partial \alpha_*}}$$
$$= \frac{h'(\beta) \int_{\alpha_*(\beta)}^d [x - \alpha^*(\beta)](x-a) w(x)\, dx}{[h'(\alpha_*(\beta)) - l'_{BA_*(\beta)}] \int_{\alpha_*(\beta)}^d (x-a)^2 w(x)\, dx} > 0$$

into (4.74), we obtain

$$\frac{dL(\beta, \alpha_*(\beta))}{d\beta} = \frac{h'(\beta)}{\int_{\alpha_*(\beta)}^d (x-a)^2 w(x)\, dx}$$
$$\times \left\{ \int_a^\beta w(x)\, dx \int_{\alpha_*(\beta)}^d (x-a)^2 w(x)\, dx \right.$$
$$+ \int_{\alpha_*(\beta)}^d w(x)\, dx \int_{\alpha_*(\beta)}^d (x-a)^2 w(x)\, dx$$
$$\left. - \left[\int_{\alpha_*(\beta)}^d (x-a) w(x)\, dx \right]^2 \right\}. \qquad (4.75)$$

The last two lines are positive by the Schwarz inequality and so is the whole expression in the curly brackets. Because $h'(\beta)$ and the denominator are positive as well, the same holds for (4.75).

We are thus led to the following conclusions. If $\hat{\beta} \geq \tilde{\beta}$ then $(\beta^*, \alpha_*(\beta_*)) = (\hat{\beta}, \hat{\beta})$ is the unique pair satisfying necessary condition (4.73) for minimizing (4.71). This gives the first statement of Lemma 17. To see that $\hat{\beta} \geq \tilde{\beta}$ coincides with (4.68), we note that (4.66) satisfies

$$\lim_{\beta \searrow a} \hat{K}(\beta) < 0 < \lim_{\beta \nearrow c} \hat{K}(\beta)$$

and
$$\hat{K}'(\beta) = h'(\beta) \int_\beta^d (x-a)w(x)\,dx > 0.$$

Therefore $\hat{K}$ is increasing and has a single zero at $\tilde{\beta}$. The same holds for $\hat{L}$ which vanishes at $\hat{\beta}$. Hence conditions $\hat{\beta} \geq \tilde{\beta}$ and (4.68) are equivalent. If $\hat{\beta} < \tilde{\beta}$, then
$$\hat{L}(\tilde{\beta}) = \lim_{\beta \nearrow \tilde{\beta}} L(\beta, \alpha_*(\beta)) > 0,$$

and we can make $L(\beta, \alpha_*(\beta))$ smaller by decreasing β. There is some $\beta_* \in [a, \tilde{\beta})$ that satisfies (4.73), because the opposite contradicts the existence of the solution. Taking $\alpha_* = \alpha_*(\beta_*)$, we see that (4.73) and (4.70) are identical, and (4.69) holds by the definition of $\alpha_*(\beta_*)$. ∎

Theorem 13 ($F \succeq_* W$) *Suppose that the density w of W and $h = f_{j:n}W$ for some $2 \leq j \leq n$ satisfy (4.29) through (4.33).*
If for unique $\hat{\beta} = \hat{\beta}(j, n) \in (0, (j-1)/(n-1))$ satisfying
$$f_{j:n}(\hat{\beta}) = 1 \tag{4.76}$$

we have
$$\int_{W^{-1}(\hat{\beta})}^d [1 - f_{j:n}W(x)](x-a)w(x)\,dx \geq 0, \tag{4.77}$$

then $\mathbb{E}_F X_{j:n} \leq \mu_F$ for all $F \succeq_* W$.
Otherwise there exists a pair (α_*, β_*), $a_W \leq \beta_* < W^{-1}((j-1)/(n-1))$, $a_W < \alpha_* \in [\beta_*, W^{-1}((j-1)/(n-1)))$, determined by equations

$$\frac{f_{j:n}W(\alpha) - f_{j:n}W(\beta)}{\alpha - a} \int_\alpha^d (x-a)^2 w(x)\,dx$$
$$+ \int_\alpha^d [f_{j:n}W(\beta) - f_{j:n}W(x)](x-a)w(x)\,dx = 0, \tag{4.78}$$

$$f_{j:n}W(\beta)\left[\int_a^\beta w(x)\,dx + \int_\alpha^d w(x)\,dx\right]$$
$$+ \frac{f_{j:n}W(\alpha) - f_{j:n}W(\beta)}{\alpha - a} \int_\alpha^d (x-a)w(x)\,dx$$
$$- F_{j:n}W(\beta) + F_{j:n}W(\alpha) - 1 = 0, \tag{4.79}$$

such that
$$\frac{\mathbb{E}_F X_{j:n} - \mu_F}{\sigma_F} \leq B = B^0_{\succeq_* W}(j, n) \tag{4.80}$$

for

$$B^2 = [f_{j:n}W(\beta_*)]^2 \left[\int_a^{\beta_*} w(x)\,dx + \int_{\alpha_*}^d w(x)\,dx \right]$$
$$+ n\frac{\binom{2j-2}{j-1}\binom{2n-2j}{n-j}}{\binom{2n-1}{n}}[F_{2j-1:2n-1}W(\alpha_*) - F_{2j-1:2n-1}W(\beta_*)]$$
$$+ 2f_{j:n}W(\beta_*)\frac{f_{j:n}W(\alpha_*) - f_{j:n}W(\beta_*)}{\alpha_* - a}\int_{\alpha_*}^d (x-a)w(x)\,dx$$
$$+ \left[\frac{f_{j:n}W(\alpha_*) - f_{j:n}W(\beta_*)}{\alpha_* - a}\right]^2 \int_{\alpha_*}^d (x-a)^2 w(x)\,dx - 1. \quad (4.81)$$

The equality in (4.80) is attained by

$$F(x) = \begin{cases} 0, & \text{if } \frac{x-\mu}{\sigma} < -\frac{1-f_{j:n}W(\beta_*)}{B}, \\ f_{j:n}^{-1}(B\frac{x-\mu}{\sigma}+1), & \text{if } -\frac{1-f_{j:n}W(\beta_*)}{B} \leq \frac{x-\mu}{\sigma} \\ & \quad < -\frac{1-f_{j:n}W(\alpha_*)}{B}, \\ W\left(a + \frac{(\alpha_*-a)[B\frac{x-\mu}{\sigma}+1-f_{j:n}W(\beta_*)]}{f_{j:n}W(\alpha_*)-f_{j:n}W(\beta_*)}\right), & \text{if } \frac{x-\mu}{\sigma} \geq -\frac{1-f_{j:n}W(\alpha_*)}{B}. \end{cases}$$
$$(4.82)$$

Trivial bounds identical with general ones for the sample minimum are consequences of

$$B^0_{\succeq_*W}(h\ 1) = B^{\nearrow}_{\succeq_*W}h\ 1 = 0$$

under (4.68), rewritten as (4.77). These apparently hold for small order statistics. Equations (4.78) and (4.79) follow from (4.69) and (4.70). Plugging in $W = U, V$ we specify (4.77) to (4.82).

Proposition 8 (decreasing density on the average) *If for given $2 \leq j \leq n-1$, and $\hat{\beta}$ defined in (4.76)*

$$1 - \hat{\beta}^2 \geq \frac{2j}{n+1}[1 - F_{j+1:n+1}(\hat{\beta})], \quad (4.83)$$

holds, then $\mathbb{E}_F X_{j:n} \leq \mu_F$.

Otherwise there are unique $0 < \beta_* < \hat{\beta}$, $\beta_* < \alpha_* < (j-1)/(n-1)$ that solve equations

$$\frac{1-\alpha^3}{3\alpha}f_{j:n}(\alpha) - \frac{(1-\alpha)^2(2+\alpha)}{6\alpha}f_{j:n}(\beta) - \frac{j[1-F_{j+1:n+1}(\alpha)]}{n+1} = 0, \quad (4.84)$$
$$\left[\beta - \frac{(1-\alpha)^2}{2\alpha}\right]f_{j:n}(\beta) + \frac{1-\alpha^2}{2\alpha}f_{j:n}(\alpha) - 1 - F_{j:n}(\beta) + F_{j:n}(\alpha) = 0, \quad (4.85)$$

and then

$$\frac{\mathbb{E}_F X_{j:n} - \mu_F}{\sigma_F} \leq B = B^0_{\succeq_*U}(j, n) \quad (4.86)$$

4.3 Distributions with Monotone Density and Failure Rate on the Average

for

$$B^2 = \left[\beta_* + \frac{(1-\alpha_*)^3}{3\alpha_*^2}\right] f_{j:n}^2(\beta_*) + \frac{1-\alpha_*^3}{3\alpha_*^2} f_{j:n}^2(\alpha_*)$$
$$- \frac{(1-\alpha_*)^2(2+\alpha_*)}{3\alpha_*^2} f_{j:n}(\beta_*) f_{j:n}(\alpha_*) - 1$$
$$+ n \frac{\binom{2j-2}{j-1}\binom{2n-2j}{n-j}}{\binom{2n-1}{n}} [F_{2j-1:2n-1}(\alpha_*) - F_{2j-1:2n-1}(\beta_*)]. \quad (4.87)$$

The equality in (4.86) holds for the location-scale family of distributions

$$F(x) = \begin{cases} 0, & \text{if } \frac{x-\mu}{\sigma} < -\frac{1-f_{j:n}(\beta_*)}{B}, \\ f_{j:n}^{-1}(B\frac{x-\mu}{\sigma} + 1), & \text{if } -\frac{1-f_{j:n}(\beta_*)}{B} \leq \frac{x-\mu}{\sigma} < -\frac{1-f_{j:n}(\alpha_*)}{B}, \\ \frac{\alpha_*[B\frac{x-\mu}{\sigma}+1-f_{j:n}(\beta_*)]}{f_{j:n}(\alpha_*)-f_{j:n}(\beta_*)}, & \text{if } -\frac{1-f_{j:n}(\alpha_*)}{B} \leq \frac{x-\mu}{\sigma} \\ & < \frac{\frac{1}{\alpha_*}f_{j:n}(\alpha_*)+(1-\frac{1}{\alpha_*})f_{j:n}(\beta_*)-1}{B}, \\ 1, & \text{if } \frac{x-\mu}{\sigma} \geq \frac{\frac{1}{\alpha_*}f_{j:n}(\alpha_*)+(1-\frac{1}{\alpha_*})f_{j:n}(\beta_*)-1}{B}. \end{cases} \quad (4.88)$$

Proposition 9 (decreasing failure rate on the average) *If for $2 \leq j \leq n$ and $\hat{\beta}$ defined in (4.76) we have*

$$(1-\hat{\beta})[1-\ln(1-\hat{\beta})] - \sum_{k=1}^{j} \frac{F_{n+1-k:n}(\hat{\beta})}{n+1-k} + \ln(1-\hat{\beta}) F_{n+1-j:n}(\hat{\beta}) \geq 0, \quad (4.89)$$

then $E_F X_{j:n} \leq \mu_F$.

Otherwise there exist $0 < \beta_* < -\ln(1-\hat{\beta})$, $\beta_* < \alpha_* < \ln(n-1)/(n-j)$ *solving equations*

$$\left(\alpha + 2 + \frac{2}{\alpha}\right) e^{-\alpha} f_{j:n}V(\alpha) - \left(1 + \frac{2}{\alpha}\right) e^{-\alpha} f_{j:n}V(\beta)$$
$$- \sum_{k=1}^{j} \frac{1}{n+1-k} F_{k:n}V(\alpha) - \alpha F_{j:n}V(\alpha) = 0, \quad (4.90)$$

$$\left(1 - e^{-\beta} - \frac{e^{-\alpha}}{\alpha}\right) f_{j:n}V(\beta) + \left(1 + \frac{1}{\alpha}\right) e^{-\alpha} f_{j:n}V(\alpha)$$
$$-1 + F_{j:n}V(\beta) - F_{j:n}V(\alpha) = 0, \quad (4.91)$$

such that

$$\frac{E_F X_{j:n} - \mu_F}{\sigma_F} \leq B = B^0_{\succeq_* V}(j,n), \quad (4.92)$$

where

78   4. Order Statistics of Independent Samples

$$\begin{aligned}
B^2 &= \left(1 - e^{-\beta_*} + 2\alpha_*^{-2} e^{-\alpha_*}\right) [f_{j:n} V(\beta_*)]^2 \\
&\quad - 2(2+\alpha_*)\alpha_*^{-2} e^{-\alpha_*} f_{j:n} V(\beta_*) f_{j:n} V(\alpha_*) \\
&\quad + (2+2\alpha_* + \alpha_*^2)\alpha_*^{-2} e^{-\alpha_*} [f_{j:n} V(\alpha_*)]^2 - 1 \\
&\quad + n \frac{\binom{2j-2}{j-1}\binom{2n-2j}{n-j}}{\binom{2n-1}{n}} [F_{2j-1:2n-1} V(\beta_*) - F_{2j-1:2n-1} V(\alpha_*)]. \quad (4.93)
\end{aligned}$$

Bound (4.92) becomes the equality for

$$F(x) = \begin{cases} 0, & \text{if } \frac{x-\mu}{\sigma} < -\frac{1-f_{j:n} V(\beta_*)}{B}, \\ f_{j:n}^{-1}(B\frac{x-\mu}{\sigma} + 1), & \text{if } -\frac{1-f_{j:n} V(\beta_*)}{B} \leq \frac{x-\mu}{\sigma} \\ & \qquad < -\frac{1-f_{j:n} V(\alpha_*)}{B}, \\ 1 - \exp\left(-\frac{\alpha_*[B\frac{x-\mu}{\sigma}+1 - f_{j:n} V(\beta_*)]}{f_{j:n} V(\alpha_*) - f_{j:n} V(\beta_*)}\right), & \text{if } \frac{x-\mu}{\sigma} \geq -\frac{1-f_{j:n} V(\alpha_*)}{B}. \end{cases}$$
(4.94)

Distribution functions (4.50) and (4.58) are similar to (4.88) and (4.94), respectively. The essential difference is that the latter ones have jumps of height $\beta_*, V(\beta_*)$, respectively. If

$$\mu_F = \frac{[1 - f_{j:n} V(\beta_*)] \sigma_F}{B^0_{\succeq_* V}(j,n)}, \quad (4.95)$$

then (4.94) is actually a DFRA life distribution starting at 0.

Under assumptions of Proposition 8, the sample maximum attains general bound (4.2), because the density of (4.3) is decreasing. Surprisingly, general bounds (4.11) and (4.4) are also attained for arbitrary order statistics from the samples with increasing density on the average. The same holds for the distribution of the narrower class of IFRA distributions, and, more generally, for families of distributions satisfying $F \preceq_* W$, when conditions (4.29) to (4.33) hold. This is a consequence of the fact that nondecreasing functions $P^{\nearrow} f_{j:n} W$, $2 \leq j \leq n$ (see (4.1) and (4.8)) may be approximated in $L^2([a_W, d_W), w(x)dx)$ with any desired accuracy by sequences h_k, $k \geq 1$, of antistarshaped nondecreasing functions starting from sufficiently low level $h_k(a)$. For example, we can take

$$h_k(x) = \min\{\alpha_k(x - \beta_k), P^{\nearrow} f_{j:n} W(x)\},$$

where $\beta_k \searrow a$, and α_k are sufficiently large. Alternatively, we can also analyze the norm convergence of $h_k W^{-1}$ in $L^2([0,1), dx)$. For more reasoning details we refer the reader to Rychlik [89]. Summarizing, we have

Theorem 14 ($F \preceq_* W$) *For $2 \leq j \leq n-1$, $h = f_{j:n}W$ and $w = dW/dx$ satisfying (4.29) to (4.33), general bounds (4.11) are attained in the limit by sequences of absolutely continuous distribution functions $F_k \preceq_* W$ whose quantile functions tend in $L^2([0,1), dx)$ to that of (4.12). In particular, the general bounds cannot be improved in the classes of distributions with increasing density and failure rate on the average. Analogous conclusions hold for the sample extremes.*

4.4 Symmetric Unimodal Distributions

For $j \leq (n+1)/2$, we have $E_F X_{j:n} \leq \mu_F$ for all $F \succeq_s U$ which is guaranteed by the symmetry assumption only. Otherwise the bounds for standardized expectations of order statistics are nontrivial, because

$$\frac{E_U X_{j:n} - \mu_U}{\sigma_U} = \sqrt{3}\left(\frac{2j}{n+1} - 1\right) > 0.$$

In the case $j = n$, the equality in (4.21) holds for a symmetric unimodal distribution. In the remaining cases $(n+1)/2 < j < n$, we obtain the optimal bounds by projecting

$$s_{j:n}(x) = f_{j:n}(x) - f_{n+1-j:n}(x), \quad 1/2 \leq x < 1,$$

onto $\mathcal{C}^+_{\succeq_c 2U-1}$. Applying Lemma 12 for verification of (4.29) through (4.33), we are in a position to describe the shape of projections by means of Lemma 13. It immediately follows that for $(n+1)/2 < j \leq \min\{n-2, [n+(3n-5)^{1/2}]/2\}$ and $j = n-1 \leq 6$, the projection is linear, and the respective bounds are attained by uniform samples. In fact, the uniform distributions provide the optimal bounds for a wider range of order statistics. The necessary and sufficient conditions are presented in Theorem 15.

Theorem 15 (symmetric unimodal distributions) *For $(n+1)/2 < j \leq n-1$, put*

$$K^s_U(x) = K_U(x) - K^-_U(x), \quad (4.96)$$
$$L^s_U(x) = L_U(x) - L^-_U(x), \quad (4.97)$$

for $1/2 \leq x < 1$, where K_U and L_U are defined in (4.48) and (4.49), respectively, and K^-_U and L^-_U are respective modifications of (4.48) and (4.49) that consist in replacing j by $n+1-j$.

If L^s_U is positive on a right neighborhood of $1/2$, then

$$\frac{E_F X_{j:n} - \mu_F}{\sigma_F} \leq \sqrt{3}\left(\frac{2j}{n+1} - 1\right), \quad (4.98)$$

where the equality holds for F uniformly distributed on $[\mu - \sqrt{3}\sigma, \mu + \sqrt{3}\sigma]$.

4. Order Statistics of Independent Samples

Otherwise, under notation (4.34) we obtain

$$\frac{E_F X_{j:n} - \mu(F)}{\sigma(F)} \leq B = \frac{\|(s_{j:n})_{\alpha_*\beta_*}\|}{\sqrt{2}}, \qquad (4.99)$$

where

$$\|(s_{j:n})_{\alpha_*\beta_*}\|^2 = n \frac{\binom{2j-2}{j-1}\binom{2n-2j}{n-j}}{\binom{2n-1}{n}} \left[F_{2j-1:2n-1}(\beta_*) + F_{2n-2j+1:2n-1}(\beta_*) \right.$$
$$\left. - F_{2j-1:2n-1}\left(\frac{1}{2}\right) - F_{2n-2j+1:2n-1}\left(\frac{1}{2}\right) \right]$$
$$- 2n \frac{\binom{n-1}{j-1}^2}{\binom{2n-1}{n}} \left[F_{n:2n-1}(\beta_*) - \frac{1}{2} \right] + (1-\beta_*) s^2_{j:n}(\beta_*)$$
$$+ \alpha_*(1-\beta_*)^2 s_{j:n}(\beta_*) + \frac{\alpha_*^2(1-\beta_*)^3}{3}, \qquad (4.100)$$

$$\alpha_* = \alpha_*(\beta_*) = \frac{3j[F_{n+2-j:n+1}(\beta_*) - F_{j+1:n+1}(\beta_*)]}{(1-\beta_*)^3(n+1)}$$
$$- \frac{3\beta_*[F_{n+1-j:n}(\beta_*) - F_{j:n}(\beta_*)]}{(1-\beta_*)^3} - \frac{3s_{j:n}(\beta_*)}{2(1-\beta_*)}, \qquad (4.101)$$

$$\beta_* = \min\left\{ \beta > \frac{1}{2} : K_U^s(\beta) L_U^s(\beta) = 0 \right\}. \qquad (4.102)$$

The equality in (4.99) is achieved by

$$F(x) = \begin{cases} 0, & \text{if } \frac{x-\mu}{\sigma} \leq -\frac{s_{j:n}(\beta_*)+\alpha_*(1-\beta_*)}{\sqrt{2B}}, \\ 1 - \beta_* + \frac{s_{j:n}(\beta_*)}{\alpha_*} + \frac{\sqrt{2B}}{\alpha_*}\frac{x-\mu}{\sigma}, & \text{if } -\frac{s_{j:n}(\beta_*)+\alpha_*(1-\beta_*)}{\sqrt{2B}} \\ & \leq \frac{x-\mu}{\sigma} \leq -\frac{s_{j:n}(\beta_*)}{\sqrt{2B}}, \\ s_{j:n}^{-1}(\sqrt{2B}\frac{x-\mu}{\sigma}), & \text{if } -\frac{s_{j:n}(\beta_*)}{\sqrt{2B}} \leq \frac{x-\mu}{\sigma} \leq \frac{s_{j:n}(\beta_*)}{\sqrt{2B}}, \\ \beta_* - \frac{s_{j:n}(\beta_*)}{\alpha_*} + \frac{\sqrt{2B}}{\alpha_*}\frac{x-\mu}{\sigma}, & \text{if } \frac{s_{j:n}(\beta_*)}{\sqrt{2B}} \leq \frac{x-\mu}{\sigma} \\ & \leq \frac{s_{j:n}(\beta_*)+\alpha_*(1-\beta_*)}{\sqrt{2B}}, \\ 1, & \text{if } \frac{x-\mu}{\sigma} \geq \frac{s_{j:n}(\beta_*)+\alpha_*(1-\beta_*)}{\sqrt{2B}}. \end{cases}$$
$$(4.103)$$

Analysis similar to that in the proof of Theorem 11 is applied here. Replacing $h = f_{j:n}$ by $h = s_{j:n} = f_{j:n} - f_{n+1-j:n}$ in (4.38) and (4.42) we write $D^s(\alpha, \beta)$, $K^s(\beta)$, $L^s(\beta)$, choosing optimal slopes (4.41) for various β, and try to determine $\beta \in [1/2, b)$ that minimizes $D^s(\alpha_*(\beta), \beta)$ under condition $K^s(\beta) \geq 0$. Since $K(\beta)$ and $L(\beta)$ of (4.42) are linear operators acting on h, we have

$$\frac{dD^s(\alpha_*(\beta), \beta)}{d\beta} = 2K^s(\beta) L^s(\beta) = 2K_U^s(\beta) L_U^s(\beta) M(\beta)$$

4.4 Symmetric Unimodal Distributions

TABLE 4.1. Sharp uniform mean-variance bounds on expectations of order statistics from independent samples of size 20 for various families of distributions.

j	G	S	SUN	DDA	DFRA
10	0.56881	0	0	0.02526	0
11	0.64211	0.17773	0.08256	0.19442	0
12	0.72127	0.50155	0.24742	0.36024	0
13	0.80855	0.74687	0.41240	0.52453	0.08848
14	0.90714	0.90447	0.57730	0.69045	0.24189
15	1.02182	1.00151	0.74232	0.86310	0.41773
16	1.16054	1.08224	0.90735	1.05096	0.62534
17	1.33774	1.18095	1.07220	1.26912	0.88206
18	1.58450	1.32488	1.24648	1.54902	1.22562
19	1.98814	1.57364	1.54236	1.97731	1.76097
20	3.04243	2.26455	2.26463	3.04423	3.03006

with a positive factor $M(\beta)$ independent of j (see (4.42) and comments following Theorem 12). A thorough analysis leads us to the conclusions that $K_U^s(\beta)$ is $+-$ on $(1/2, b)$, and $L_U^s(\beta)$ is either $+$ or $-+$ on $(1/2, c)$. If $L_U^s(1/2+) > 0$, then $\beta_* = 1/2$ is the solution that implies linearity of projection and the resulting quantile function. Otherwise optimal $\beta_* > 1/2$ is the smaller of zeros of K_U^s and L_U^s. Then

$$P^+_{\succeq_c 2U-1} s_{j:n} = (s_{j:n})_{\alpha_*(\beta_*)\beta_*},$$

and (4.100), (4.101), and (4.103) are derived from general formulae by elementary calculations.

Note that (4.103) has a density symmetric about μ_F, a finite support with uniform ends, and a (principally infinite) peak at the center. The statement of Theorem 15 is weaker than those of Theorems 11 and 12 in that we were not able to determine explicitly the pairs (j, n) for which $L(a) > 0$, and the resulting optimal bounds are determined by the minimal distributions W in the class with respect to the order.

In Table 4.1 numerical evaluations of mean-variance bounds are presented for the jth order statistics, $10 \leq j \leq 20$, of i.i.d. samples of size $n = 20$, coming from general (G), symmetric (S), symmetric unimodal (SUN) populations, and those with decreasing density and failure rate on the average (DDA and DFRA, respectively). For $j \leq 9$, all bounds are trivial except for the first case. The values of the third column were presented in Gajek and Rychlik [33, Table III]. Numerical bounds for the DDA and DFRA samples of size 15 can be found in Rychlik [89].

4.5 Bias of Quantile Estimates

Order statistic $X_{j:n}$ is the most natural nonparametric estimate of quantile $F^{-1}(p)$ if j/n is close to p, although some more sophisticated smooth estimates based on kernels and Bernstein polynomial estimates were also constructed (see, e.g., Sheather and Marron [97], Huang and Brill [40], and Cheng [21]). If $j(n)/n \to p$ and $F^{-1}(p)$ is unique, then $X_{j(n):n} \to F^{-1}(p)$ almost surely. Optimality of order statistics in quantile estimation for finite samples under various criteria was proved by Zieliński [103, 105, 106]. Here we employ the projection method for gauging the bias $E_F X_{j:n} - F^{-1}(p)$ of quantile estimation by sample quantiles. For convenience we assume that $j/n = p$ precisely. We first consider general distribution functions. Since both the upper and lower bias deviations are significant here, we present the lower bounds for the difference as well.

Theorem 16 (general distributions) *For arbitrary* $1 \leq j < n < \infty$,

$$\frac{E_F X_{j:n} - F^{-1}(p)}{\sigma_F} \leq \bar{B}^{\nearrow}(j, n, p) = \frac{1 - F_{j:n}(p)}{[p(1-p)]^{1/2}}. \tag{4.104}$$

For $j = 2$ *and* $n = 3, 4, 5, 6$, *and* $3 \leq j < n < \infty$,

$$\frac{E_F X_{j:n} - F^{-1}(p)}{\sigma_F} \geq -\underline{B}^{\nearrow}(j, n, p) = -\frac{F_{j:n}(p)}{[p(1-p)]^{1/2}}. \tag{4.105}$$

Both (4.104) and (4.105) are attained by the (limiting) two-point distribution

$$P\left(X = \mu - \sigma\left[\frac{1-p}{p}\right]^{1/2}\right) = p = 1 - P\left(X = \mu + \sigma\left[\frac{p}{1-p}\right]^{1/2}\right). \tag{4.106}$$

For $j = 1 < n < \infty$,

$$\frac{E_F X_{1:n} - F^{-1}(p)}{\sigma_F} \geq -\underline{B} = -\underline{B}^{\nearrow}(1, n, p), \tag{4.107}$$

where

$$\underline{B}^2 = \frac{n^2}{2n-1} F_{1:2n-1}\left(\frac{1}{n}\right) + \frac{n}{n-1} F_{1:n}^2\left(\frac{1}{n}\right). \tag{4.108}$$

The equality in (4.107) holds for

$$F(x) = \begin{cases} 0, & \text{if } \frac{x-\mu}{\sigma} < -\frac{n}{\underline{B}}, \\ 1 - \left(-\frac{\underline{B}}{n}\frac{x-\mu}{\sigma}\right)^{1/(n-1)}, & \text{if } -\frac{n}{\underline{B}} \leq \frac{x-\mu}{\sigma} \leq -\frac{(n-1)^{n-1}}{\underline{B} n^{n-2}}, \\ \frac{1}{n}, & \text{if } -\frac{(n-1)^{n-1}}{\underline{B} n^{n-2}} \leq \frac{x-\mu}{\sigma} < \frac{n F_{1:n}(1/n)}{(n-1)\underline{B}}, \\ 1, & \text{if } \frac{x-\mu}{\sigma} \geq \frac{n F_{1:n}(1/n)}{(n-1)\underline{B}}. \end{cases} \tag{4.109}$$

For $j = 2$ and $n \geq 7$, we have

$$\frac{E_F X_{2:n} - F^{-1}(p)}{\sigma_F} \geq -\underline{B} = -\underline{B}^{\nearrow}(2, n, p), \qquad (4.110)$$

where

$$\begin{aligned}\underline{B}^2 &= \frac{(n-1)n^2}{(2n-3)(2n-1)} \left[F_{3:2n-1}\left(\frac{2}{n}\right) - F_{3:2n-1}(\alpha_*) \right] \\ &\quad + \alpha_* f_{2:n}^2(\alpha_*) + \frac{n}{n-2} F_{2:n}^2\left(\frac{2}{n}\right), \end{aligned} \qquad (4.111)$$

and $\alpha_* \in (1/(n-1), 2/n)$ is uniquely defined by

$$[(n-1)^2 x^2 + (n-1)x + 1](1-x)^{n-2} = 1. \qquad (4.112)$$

Bound (4.110) is attained by the distribution function

$$F(x) = \begin{cases} 0, & \text{if } \frac{x-\mu}{\sigma} < -\frac{f_{2:n}(\alpha_*)}{\underline{B}}, \\ f_{2:n}^{-1}(-\underline{B}\frac{x-\mu}{\sigma}), & \text{if } -\frac{f_{2:n}(\alpha_*)}{\underline{B}} \leq \frac{x-\mu}{\sigma} \leq -\frac{f_{2:n}(2/n)}{\underline{B}}, \\ \frac{2}{n}, & \text{if } -\frac{f_{2:n}(2/n)}{\underline{B}} \leq \frac{x-\mu}{\sigma} < \frac{nF_{2:n}(2/n)}{(n-2)\underline{B}}, \\ 1, & \text{if } \frac{x-\mu}{\sigma} \geq \frac{nF_{2:n}(2/n)}{(n-2)\underline{B}}. \end{cases} \qquad (4.113)$$

Note that

$$\bar{B}(j, n, p) - \underline{B}(j, n, p) \geq [p(1-p)]^{-1/2}. \qquad (4.114)$$

Precisely, we have equality in (4.114) under conditions of (4.105), and strict inequality holds only in the exceptional cases treated in the last two statements. This implies that the bias oscillation presumably does not depend on the sample size, is smaller for the central quantiles, and increases to infinity on the tails. The only positive effect of the sample increase is that the absolute deviation of the bias decreases to $1/[4p(1-p)]^{1/2}$, because $F_{j:n}(p) \to 1/2$ by the de Moivre–Laplace theorem. However, for no quantile the bias oscillation tends to zero. This is obvious, because we take into account the distributions with nonunique pth quantiles (estimation problems in such cases are discussed in Feldman and Tucker [28]). Removing such distributions would not help here either, because the rate of convergence of $X_{j:n}$ to $F^{-1}(p)$ depends on the slope increase of F at $F^{-1}(p)$, and it is impossible to determine uniform rates without imposing conditions on the slope (cf., e.g., Zieliński [104]). In fact, all the bounds of Theorem 16 are optimal for the class of strictly increasing F. However, the conclusions on attainability should be formulated more carefully then.

The proof of Theorem 16 is based on the representation

$$E_F X_{j:n} - F^{-1}(p) = \int_0^1 F^{-1}(x)(F_{j:n} - \mathbf{1}_{[p,1)})(dx)$$

and Lemma 3. The upper bound (4.104) is the L^2-norm of the derivative of the greatest convex minorant of $F_{j:n} - \mathbf{1}_{[p,1)}$, and the centered quantile function of (4.106) is proportional to the derivative. Since $F_{j:n} - \mathbf{1}_{[p,1)}$ starts from 0, increases on $[0, p)$, jumps down to $F_{j:n}(p) - 1 < 0$, increases to 0 on $[p, 1)$, and is concave on $[p, 1)$, its greatest convex minorant has two linear pieces on $[0, p]$ and $[p, 1]$ with slopes $-[1 - F_{j:n}(p)]/p$ and $[1 - F_{j:n}(p)]/(1-p)$, respectively. An easy computation leads us to the final claim.

In order to get the lower bounds it suffices to find the greatest convex minorant for $\mathbf{1}_{[p,1)} - F_{j:n}$. Distribution function $F_{1:n}$ is concave, and therefore the greatest convex minorant of $\mathbf{1}_{[p,1)}(x) - F_{j:n}(x)$ coincides with $-F_{1:n}(x)$ and the straight line $F_{1:n}(p)(x-1)/(1-p)$ on $[0, p)$ and $[p, 1)$, respectively. For $2 \leq j < n$, the problem is that function $\mathbf{1}_{[p,1)}(x) - F_{j:n}(x) = -F_{j:n}(x)$ for $0 \leq x \leq p = j/n$ is decreasing concave-convex with the inflection point $(j-1)/(n-1) < p$. Two cases are possible: either the straight line joining $(0, -F_{j:n}(0)) = (0, 0)$ with $(p, -F_{j:n}(p))$ lies entirely below the graph of $-F_{j:n}$, or the line crosses the graph there. In the former case, the line becomes a part of the greatest convex minorant. In the latter one, the minorant consists of the line passing through $(0, 0)$, and tangent to the graph at a point $\alpha_* \in ((j-1)/(n-1), p)$, and $-F_{j:n}(x)$ itself on $[\alpha_*, p]$. In both cases, the line passing through $(p, -F_{j:n}(p))$ and $(1, \mathbf{1}_{[p,1)}(1) - F_{j:n}(1)) = (1, 0)$ is the remaining part of the minorant. The problem of settling which case actually holds is equivalent to checking the sign of the expression

$$b(j,n) = F_{j:n}\left(\frac{j}{n}\right) - \frac{j}{n} f_{j:n}\left(\frac{j}{n}\right). \tag{4.115}$$

We have $b(j, n) \leq 0$ in the first case, which leads to (4.105), and (4.108) holds otherwise. A thorough study carried out in Okolewski and Rychlik [65] shows that (4.115) is positive for $j = 2$ with $n \geq 7$ only. This is based on analysis of sign changes of sequences defined as integrals of a fixed function with sequences of totally positive functions (see Karlin [43] and [44, Chapter 1] for details). We preserved the dual notation j/n and p in Theorem 16 on purpose. The reason is that the formulae remain true if $p \approx j/n$ as well. For instance, both (4.104) and (4.105) hold for all j, n with $0 < p = (j-1)/(n-1) < 1$. In fact, the former is true for all $1 < j < n$ with $p \geq (j-1)/(n-1)$ and so is the latter for all $1 < j < n$ with $p \leq (j-1)/(n-1)$.

Now we restrict ourselves to the distributions following a given W in the convex and star orders. We present the upper bounds only. As above, all are true for arbitrary j, n with $p \geq (j-1)/(n-1)$. The auxiliary dual problem to solve is to maximize the functional

$$E_F X_{j:n} - F^{-1}(p)$$
$$= \lim_{q \searrow p} \int_0^1 F^{-1}(x) \left[f_{j:n}(x) - \frac{1}{q-p} \mathbf{1}_{[p,q)}(x) \right] dx$$

$$= \lim_{c \searrow b} \int_a^d F^{-1}W(x) \left[f_{j:n}W(x) - \frac{\mathbf{1}_{[b,c)}(x)}{W(c) - W(b)} \right] w(x)dx \quad (4.116)$$

over $\mathcal{C}^0_{\succeq_c W}$ and $\mathcal{C}^0_{\succeq_* W}$. For convenience, we replace the constant subtrahend in the brackets by a variable one to make the difference constant. This does not affect the limit

$$\mathbb{E}_F X_{j:n} - F^{-1}(p)$$
$$= \lim_{c \searrow b} \int_a^d F^{-1}W(x) \Big[f_{j:n}W(x)\mathbf{1}_{[a,d)\setminus[b,c)}(x)$$
$$- \frac{1 - F_{j:n}W(c) + F_{j:n}W(b)}{W(c) - W(b)} \mathbf{1}_{[b,c)}(x) \Big] w(x)\, dx. \quad (4.117)$$

We are interested in projecting $h_{j:n,b,c}(x)$ which denotes the expression in the square brackets.

Lemma 18 *If $h \in L^2([a_W, d_W), w(x)dx)$ is nonnegative on $[a, b)$, negative constant on $[b, c)$, and nonnegative nonincreasing on $[c, d)$ for $a < b < c < d$, and*

$$\int_a^d h(x)w(x)\, dx = 0, \quad (4.118)$$

then for every $g \in \mathcal{C}^0_{\succeq_c W}$ there exists $g_{\alpha\beta\gamma} \in \mathcal{C}^0_{\succeq_c W}$ defined as

$$g_{\alpha\beta\gamma}(x) = \alpha(x - \gamma)\mathbf{1}_{[\gamma,d)}(x) - \beta \quad (4.119)$$

with $\alpha, \beta > 0$ and $b \le \gamma \le c$ such that

$$\|g_{\alpha\beta\gamma} - h\| \le \|g - h\|.$$

Function $h_{j:n,b,c}$ is orthogonal to constants and satisfies the other assumptions of Lemma 18 if $p \ge (j-1)/(n-1)$. The orthogonality property (4.118) is important here: it allows us to restrict assumptions on $h_{j:n,b,c}$ to sign and monotonicity conditions. In other lemmas describing the shape of projections, the concavity, and convexity in some regions play a significant role as well.

PROOF OF LEMMA 18. Since g is nondecreasing and integrates to 0, the only possible constant function is equal to 0. Otherwise we have $g(a) < 0 < g(d-)$. If $g(b) \ge 0$, then

$$g_+ = \max\{g, 0\} \in \mathcal{C}^{\nearrow}_{\succeq_c W}$$

lies closer to h, and, by Lemma 2, so does

$$\mathcal{C}^0_{\succeq_c W} \ni g^0_+ = g_+ - (g_+, \mathbf{1}) < g_+.$$

Repetitive application of the procedure allows us to restrict ourselves to functions satisfying $0 > g(b) = -\beta$, say. Similar arguments, with g_+ replaced by $g_{h(b)} = \max\{g, h(b)\}$ excludes the functions for which $g(a) < h(b) = \inf_{a \leq x < d} h(x)$. We are therefore in a position to assume

$$h(b) \leq g(a) \leq g(b) \leq 0,$$

and define
$$\delta = \inf\{x \geq : g(x) \geq h(x)\}, \qquad (4.120)$$

setting $\delta = d$ if $d < \infty$ and $g(d-) \leq h(d-)$.

If $\delta > c$, the straight line $l_{c\delta}$ secant to g at c and δ has a positive slope

$$\alpha = \frac{g(\delta) - g(c)}{\delta - c},$$

and runs above g in $[c, \delta]$ and below g elsewhere. In particular, we have $g(b) \leq l_{c\delta}(b)$. It follows that $l_{c\delta}$ crosses the constant function

$$l_b(x) \equiv g(b) = -\beta$$

at a $\gamma \in [b, c)$. We easily see that the broken line

$$g_{\alpha\beta\gamma}(x) = \max\{l_{c\delta}(x), -\beta\} = \alpha(x - \gamma)\mathbf{1}_{[\gamma, d)}(x) - \beta$$

satisfies

$$g(x) \leq g_{\alpha\beta\gamma}(x) \leq h(x) \quad \text{if} \quad x \in (a, b) \cup (c, \delta), \qquad (4.121)$$
$$h(x) \leq g_{\alpha\beta\gamma}(x) \leq g(x) \quad \text{if} \quad x \in (b, c) \cup (\delta, d), \qquad (4.122)$$

which is the desired conclusion.

If $\delta = c$ we replace $l_{c\delta}$ by the line l_c tangent to g at c. It runs beneath g, and crosses l_b at some $\gamma \in [b, c)$. Its truncation at the level $l_b(\gamma) = -\beta$ has form (4.119), and satisfies (4.121) and (4.122) with $(c, \delta) = \emptyset$. ∎

Lemma 19 *Under the hypotheses and notation of Lemma 18, we have*

$$P^0_{\succeq c W} h(x) = g_{\alpha_* \beta_* \gamma_*}(x)$$

for some $\gamma_* \in [b, c]$,

$$\alpha_* = \alpha_*(\gamma_*) = \frac{\int_{\gamma_*}^d (x - \gamma_*) h(x) w(x) dx}{\int_{\gamma_*}^d (x - \gamma_*)^2 w(x) dx - \left[\int_{\gamma_*}^d (x - \gamma_*) w(x) dx\right]^2}, \qquad (4.123)$$

$$\beta_* = \beta_*(\gamma_*) = \alpha_*(\gamma_*) \int_{\gamma_*}^d (x - \gamma_*) w(x) dx. \qquad (4.124)$$

4.5 Bias of Quantile Estimates

PROOF. For fixed α, γ, the function

$$D(\alpha, \beta, \gamma) = \int_a^d [h(x) - \alpha(x - \gamma)\mathbf{1}_{[\gamma,d)}(x) + \beta]^2 w(x)\, dx$$

is minimized with respect to translation parameter β at

$$\begin{aligned}\beta_*(\alpha, \gamma) &= \int_a^d [\alpha(x-\gamma)\mathbf{1}_{[\gamma,d)}(x) - h(x)]w(x)\, dx \\ &= \alpha \int_\gamma^d (x-\gamma) w(x)\, dx\end{aligned}$$

(cf. (4.124)). Furthermore, for fixed $\gamma \in [b,c]$,

$$\begin{aligned}&D(\alpha, \beta_*(\alpha, \gamma), \gamma) \\ &= \int_a^d \left\{ h(x) - \alpha\left[(x-\gamma)\mathbf{1}_{[\gamma,d)}(x) - \int_\gamma^d (y-\gamma)w(y)\, dy\right]\right\}^2 w(x)\, dx\end{aligned}$$

is a convex quadratic function in α, with the global minimum at $\alpha_*(\gamma)$, defined as in (4.123). It is obvious that

$$\gamma \mapsto D(\alpha_*(\gamma), \beta_*(\alpha_*, \gamma), \gamma)$$

is a continuous function that attains its minimum at some γ_*. Observe that the numerator of the right-hand side of (4.123) is positive. By the Schwarz inequality with condition $\int_\gamma^d w(x)\, dx < 1$, the denominator is positive as well. Therefore $\alpha_*(\gamma_*) > 0$, which together with (4.124) yields $\beta_* = \beta_*(\alpha_*(\gamma_*), \gamma_*) > 0$. The solution to our minimization problem is actually a convex nondecreasing function. ■

If $c \searrow b$, so does γ_*. For $h = h_{j:n,b,c}$ we have

$$\begin{aligned}0 &< -\int_{\gamma_*}^c (x-\gamma_*)h(x)w(x)\, dx \\ &= \frac{1 - F_{j:n}W(c) + F_{j:n}W(b)}{W(c) - W(b)} \int_{\gamma_*}^c (x - \gamma_*)w(x)\, dx \\ &< (c - \gamma_*)[1 - F_{j:n}W(c) + F_{j:n}W(b)] \to 0 \quad (4.125)\end{aligned}$$

and so

$$\begin{aligned}\alpha_*(\gamma_*) \to \alpha_*(b) &= \frac{\int_b^d (x-b) f_{j:n}W(x)w(x)\, dx}{\int_b^d (x-b)^2 w(x)\, dx - \left[\int_b^d (x-b)w(x)\, dx\right]^2}, \\ \beta_*(\gamma_*) \to \beta_*(b) &= \alpha_*(b) \int_b^d (x-b)w(x)\, dx,\end{aligned}$$

88   4. Order Statistics of Independent Samples

as $c \searrow b$. We easily see that

$$\mathrm{E}_F X_{j:n} - F^{-1}(p)$$
$$\leq \int_a^d [F^{-1}W(x) - \mu_F] g_{\alpha_*(b)\beta_*(b)b}(x)\,dx$$
$$\leq \frac{\int_b^d (x-b) f_{j:n}W(x) w(x)\,dx}{\left\{\int_b^d (x-b)^2 w(x)\,dx - \left[\int_b^d (x-b) w(x)\,dx\right]^2\right\}^{1/2}} \sigma_F, \quad (4.126)$$

which becomes the equality if

$$\frac{F^{-1}W(x) - \mu_F}{\sigma_F} = \frac{(x-b)\mathbf{1}_{[b,d)}(x) - \int_b^d (x-b) w(x)\,dx}{\left\{\int_b^d (x-b)^2 w(x)\,dx - \left[\int_b^d (x-b) w(x)\,dx\right]^2\right\}^{1/2}}.$$

Theorem 17 ($F \succeq_c W$) Set

$$\hat{\eta} = \hat{\eta}_W(b) = \mathrm{E}_W(X-b)_+ = \int_b^d (x-b) w(x)\,dx, \quad (4.127)$$

$$\hat{\eta}_{W_{j:n}}(b) = \mathrm{E}_W(X_{j:n}-b)_+$$
$$= \int_b^d (x-b) f_{j:n}W(x) w(x)\,dx, \quad (4.128)$$

$$\hat{\vartheta}^2 = \hat{\vartheta}_W^2(b) = \mathrm{Var}_W(X-b)_+$$
$$= \int_b^d (x-b) w(x)\,dx - \left[\int_b^d (x-b) w(x)\,dx\right]^2. \quad (4.129)$$

If $(j-1)/(n-1) \leq p \in (0,1)$, then for $b = W^{-1}(p)$

$$\frac{\mathrm{E}_F X_{j:n} - F^{-1}(p)}{\sigma_F} \leq \frac{\hat{\eta}_{W_{j:n}}(b)}{\hat{\vartheta}_W(b)}, \quad (4.130)$$

and we get the equality here if

$$F(x) = W\!\left(W^{-1}(p) + \hat{\eta} + \hat{\vartheta}\,\frac{x-\mu}{\sigma}\right)\mathbf{1}_{[\mu-\sigma\hat{\eta}/\hat{\vartheta},+\infty)}(x). \quad (4.131)$$

Note that

$$\hat{\eta}_W(b) = \eta_W(b,b),$$
$$\hat{\vartheta}_W(b) = \vartheta_W(b,b),$$

where the right-hand sides were defined in (3.59) and (3.60).

Proposition 10 (decreasing density) *For* $(j-1)/(n-1) \leq p \in (0,1)$, *we have*

$$\frac{E_F X_{j:n} - F^{-1}(p)}{\sigma_F} \leq 2\sqrt{3} \frac{\frac{j}{n+1}[1 - F_{j+1:n+1}(p)] - p[1 - F_{j:n}(p)]}{(1-p)^{3/2}(1+3p)^{1/2}}. \quad (4.132)$$

Bound (4.132) *is attained by the mixture of the atom at* $\mu - \sigma[3(1-p)/(1+3p)]^{1/2}$ *with weight* p, *and the uniform distribution on* $[\mu - \sigma[3(1-p)/(1+3p)]^{1/2}, \mu + \sigma(1+p)\{3/[(1-p)(1+3p)]\}^{1/2}]$ *with weight* $1-p$.

Proposition 11 (decreasing failure rate) *For* $(j-1)/(n-1) \leq p \in (0,1)$, *the inequality*

$$\frac{E_F X_{j:n} - F^{-1}(p)}{\sigma_F} \leq \frac{1}{(1-p^2)^{1/2}} \sum_{k=1}^{j} \frac{1 - F_{k:n}(p)}{n+1-k} \quad (4.133)$$

is tight and attained by the mixture of the atom at $\mu - \sigma[(1-p)/(1+p)]^{1/2}$, *and the exponential distribution with scale parameter* $\sigma/(1-p^2)^{1/2}$ *starting from the atom, with respective weights* p *and* $1-p$.

For $F \succeq_* W$, the solution has a similar and even simpler form.

Lemma 20 *Under the assumptions of Lemma 18 for every* $g \in C^0_{\succeq_* W}$ *there exists* $g_{\alpha\beta} \in C^0_{\succeq_* W}$ *defined as*

$$g_{\alpha\beta}(x) = \alpha(x-a)\mathbf{1}_{[c,d)}(x) - \beta \quad (4.134)$$

with $\alpha, \beta > 0$ *such that*

$$\|g_{\alpha\beta} - h\| \leq \|g - h\|.$$

PROOF. By arguments analogous to those of Lemma 18, it suffices to confine ourselves to the family of starshaped nondecreasing functions g such that $g(b) = -\beta$ for some $\beta \in [h(b), 0]$. For all $x \leq b$ we have

$$g(x) \leq -\beta \leq 0 \leq h(x).$$

This means that the approximation of h is improved by

$$g_{-\beta} = \max\{g, -\beta\} \in C^{\nearrow}_{\succeq_* W}.$$

Recalling (4.120), we define (4.134) with the slope parameter

$$\alpha = \frac{g(\delta) + \beta}{\delta - a} > 0.$$

Then the function (4.134) is the smallest possible one for arguments $x \in (b, \delta)$, and the greatest one for $x \in (\delta, d)$ among all starshaped functions starting from $(a, -\beta)$ and passing through $(\delta, g(\delta))$. We have thus proved (4.121) and (4.122), with (4.119) replaced by (4.134), which is precisely the assertion of the lemma. ∎

Lemma 21 *Under the assumptions of Lemma 18 and notation of Lemma 20, we have*
$$P^0_{\succeq_* W} h(x) = g_{\alpha_* \beta_*}(x)$$
for

$$\alpha_* = \frac{\int_c^d (x-a) h(x) w(x)\, dx}{\int_c^d (x-a)^2 w(x)\, dx - \left[\int_c^d (x-a) w(x)\, dx\right]^2}, \qquad (4.135)$$

$$\beta_* = \alpha_* \int_c^d (x-a) w(x)\, dx. \qquad (4.136)$$

The proof is similar to that of Lemma 19. We first find the solution (4.136) to the minimization problem of $\|g_{\alpha\beta} - h\|^2$ with respect to β. Then we look for α_* minimizing $\|g_{\alpha\beta_*(\alpha)\gamma} - h\|^2$ and find (4.135), which is positive by arguments analogous to those presented in the last part of the proof of Lemma 19.

Taking $h(x) = f_{j:n} W(x)$ for $x \geq c$, and passing to the limit, we get

$$\|g_{\alpha_* \beta_*}\| = \frac{\int_b^d (x-a) f_{j:n} W(x) w(x)\, dx}{\left\{\int_b^d (x-a)^2 w(x)\, dx - \left[\int_b^d (x-a) w(x)\, dx\right]^2\right\}^{1/2}}, \qquad (4.137)$$

and the normalized function $g_{\alpha_* \beta_*}/\|g_{\alpha_* \beta_*}\|$ identical to (3.58). This implies that the distribution functions which attain the bounds in Theorem 18, and Propositions 12 and 13 coincide with those attaining the bounds of Theorem 6, and Propositions 5 and 6, respectively.

Theorem 18 ($F \succeq_* W$) *Under the notation (3.60), $b = W^{-1}(p)$, and*

$$\eta_{W_{j:n}}(a, b) = E_W (X_{j:n} - a) \mathbf{1}_{[b,d)}(X_{j:n})$$
$$= \int_b^d (x-a) f_{j:n} W(x) w(x)\, dx, \qquad (4.138)$$

and for $(j-1)/(n-1) \leq p \in (0,1)$ the following inequality is sharp

$$\frac{E_F X_{j:n} - F^{-1}(p)}{\sigma_F} \leq \bar{B}^0_{\succeq_* W}(j, n, p) = \frac{\eta_{W_{j:n}}(a, b)}{\vartheta_W(a, b)}. \qquad (4.139)$$

Proposition 12 (decreasing density on the average) *If $(j-1)/(n-1) \leq p \in (0,1)$, then the following inequality is sharp*

$$\frac{E_F X_{j:n} - F^{-1}(p)}{\sigma_F} \leq \bar{B}^0_{\succeq_* U}(j, n, p)$$
$$= \frac{2\sqrt{3}j}{n+1} \frac{1 - F_{j+1:n+1}(p)}{\theta_U(p)}. \qquad (4.140)$$

Proposition 13 (decreasing failure rate on the average) *If we have $(j-1)/(n-1) \leq p \in (0,1)$, then the following inequality is sharp*

$$\frac{E_F X_{j:n} - F^{-1}(p)}{\sigma_F} \leq \bar{B}^0_{\succeq_* V}(j,n,p)$$

$$= \frac{\sum_{k=1}^{j} \frac{1-F_{k:n}(p)}{n+1-k} - \ln(1-p)[1-F_{j:n}(p)]}{\theta_V(p)}. \quad (4.141)$$

Notations $\theta_U(p)$ and $\theta_V(p)$, that appear in (4.140) and (4.141), were defined in (3.64) and (3.67), respectively. It is worth pointing out that the upper bias deviations in large samples are significantly different for the classes of distributions determined by the convex and star orders. To see this, suppose that p is fixed and take a sequence of order statistics $X_{j:n}$, $j = j(n), n \to \infty$, such that $p \geq (j-1)/(n-1) \to p$. If $F \succeq_c W$, then (4.130) holds and, since $X_{j:n} \to b = W^{-1}(p)$ almost surely, (4.128) tends to 0. On the other hand, (4.129) does not depend on the sample size and is positive. In consequence, the bound in (4.130) tends to 0, and so do those of (4.132) and (4.133). If $F \succeq_* W$, then (4.138) tends to $(b-a)/2 > 0$, and (3.60) remains fixed positive. Therefore the right-hand sides of all (4.139) to (4.141) have positive limits.

Table 4.2 contains values of extreme upper deviations of order statistics estimates $X_{j:n}$, $1 \leq j < n = 20$, of quantiles of order $p = j/n$ for the families of general (G), decreasing density (DD) and failure rate (DFR), and decreasing density, failure rate on the average (DDA and DFRA, respectively). In general populations, the estimates are more stable for the central quantiles than for the extreme ones which confirms the theoretical analysis of bias oscillation (cf. (4.114)). Otherwise the deviations are increasing in j. It follows from the fact that all the families are more concentrated on the left, and the upper quantiles are more dispersed and thus more difficult to be estimated.

4.6 Open Problems

1. Natural questions are the mean-variance bounds on expectations of single order statistics for populations determined by the convex order. In particular, what are the evaluations for distributions with monotone density and failure rate?

2. General second moment bounds are presented in (4.18), (4.14), and (4.15). Analogous results for distributions with decreasing density and failure rate can be found in Section 4.2, but the problem is unsolved when either density of failure rate is increasing. More generally, we ask for the second moment bounds for families of life distributions determined by the convex order. Likewise, what are the respective bounds for distributions determined by the star order?

TABLE 4.2. Sharp uniform variance bounds on upper bias deviations of estimators $X_{j:20}$ of pth quantiles, $p = j/20$, for various families of distributions (independent case).

j	G	DD	DFR	DDA	DFRA
1	1.64485	0.05658	0.01795	0.11742	0.03627
2	1.30582	0.08034	0.02683	0.20820	0.06764
3	1.13394	0.09763	0.03406	0.28585	0.09838
4	1.02862	0.11226	0.04069	0.35251	0.12916
5	0.95804	0.12575	0.04718	0.40974	0.16022
6	0.90860	0.13894	0.05379	0.45924	0.19170
7	0.87348	0.15240	0.06075	0.50272	0.22379
8	0.84894	0.16660	0.06828	0.54180	0.25671
9	0.83279	0.18203	0.07660	0.57798	0.29076
10	0.82380	0.19919	0.08603	0.61264	0.32638
11	0.82139	0.21876	0.09694	0.64709	0.36418
12	0.82548	0.24158	0.10989	0.68269	0.40502
13	0.83648	0.26888	0.12569	0.72092	0.45011
14	0.85539	0.30247	0.14557	0.76365	0.50125
15	0.88410	0.34523	0.17163	0.81339	0.56122
16	0.92588	0.40204	0.20756	0.87393	0.63458
17	0.98657	0.48192	0.26077	0.95145	0.72938
18	1.07691	0.60355	0.34835	1.05688	0.86128
19	1.21205	0.80859	0.51815	1.20493	1.06011

3. Differences of order statistics, especially the sample range $X_{n:n} - X_{1:n}$ and *quasiranges* $X_{j:n} - X_{n+1-j:n}$, $(n+1)/2 < j < n$, are used as simple and robust estimates of dispersion of the parent population. *Spacings* $X_{j:n} - X_{j-1:n}$, $2 \leq j \leq n$, represent times between consecutive failure times of elements of composite systems. In general populations, expectations of sample range and quasiranges were precisely evaluated by Plackett [72] and Moriguti [58], respectively. The former was derived by direct application of the Schwarz inequality. For the latter, the greatest convex minorant construction was used. In both cases, the bounds are twice as great as those on the jth order statistic in the symmetric populations (see (4.21) and (4.24) through (4.26)). It is important to find evaluations for differences of order statistics, especially the quasiranges and spacings, in the restricted families of distributions considered here.

4. Completely untouched domains of potential interest are bounds on most popular L-statistics: trimmed means, some best linear unbiased estimates (so called BLUEs) in parametric models, the *Gini mean difference* $2/[n(n-1)] \sum_{j=1}^{n}(2j - n - 1)X_{j:n}$, and many others, both in general and restricted families of distributions.

5. The results of Section 4.5 should be completed by analyzing the bias of quantile estimates for dual families of distributions defined by relations $F \preceq_c W$ and $F \preceq_* W$. Evaluations of the bias in the symmetric populations, with possible additional restrictions, are not known, either. Also, some L-statistics can be studied in a further perspective. We point out here that in the problems of bias evaluation, the lower bounds are of key interest as well. On the other hand, the upper and lower bounds on $E_F X_{j:n} - E_F X_{kj:kn}$ for various $k > 1$ would provide precise evaluations of rates of convergence of quantile estimates over different classes of distributions.

6. The order statistic $X_{j:n}$ is used for estimating $F^{-1}(p)$ if j/n is close to p. This is justified by strong consistency of the estimates of the unique quantiles for which $j/n \to p$ is only needed. If n is fixed and $j/n < p < (j+1)/n$, then the $(j+1)$st order statistic is a possibly better candidate. Evaluating $E_F X_{j:n} - F^{-1}(p)$ for various j, and given n and p, we can find the order statistic whose bias is minimal in a given class of distributions, and recommend it as the most bias-robust estimate of the pth quantile. Therefore it is worth extending results of Section 4.5 to the cases $p \neq j/n$.

5
Order Statistics of Dependent Observations

Assume that $Y_1, \ldots, Y_n$ are *possibly dependent* and identically distributed. Recalling arguments of Rychlik [79], in Section 5.1 we conclude sharp bounds (2.25) and (2.27) on expectations of general L-statistics and single order statistics, respectively, depending on the common marginal distribution of the observations. Next we apply the projections of functionals defined in (2.25) and (2.27) for establishing respective moment bounds over general and restricted families of marginals. In Section 5.2 general, symmetric, and nonnegative observations are treated, and respective deterministic bounds for arbitrary samples are concluded. The results cited are from Rychlik [81], but some earlier partial solutions are also mentioned. In the remainder of Chapter 5 we confine ourselves to single order statistics. In Section 5.3 we present mean-variance and second moment bounds on the expectations of order statistics for families of parent distributions related to a given one in the convex order. The mean-variance bounds for $F \succeq_c W$ and $F \preceq_c W$ are not published elsewhere, except for DFR and IFR distributions, given in Rychlik [87]. The results for general W and $W = U$ (i.e., for the decreasing density distributions) are presented here as well. Rychlik [84] established second moment bounds for $F \succeq_c (\preceq_c) W$ with general W. The decreasing density and failure rate distributions were studied in Gajek and Rychlik [32]. Analogous results for increasing ones come from Rychlik [84]. Section 5.4 deals with mean-variance bounds for order statistics based on samples with common marginal distributions being in a star relation with a fixed one W. For ones determined by the exponential $W = V$, which includes important classes of life distributions with monotone failure rate, respective bounds were established in Rychlik [87]. Especially, it was shown that general bounds (5.13) are attained by the IFRA distributions. In fact, the claim can be extended to families of distributions defined by $F \preceq_* W$ for general W. We also write explicitly the bounds for $F \succeq_* U$ that have decreasing densities on the average. The mean-variance bounds for order statistics with symmetric unimodal and U-shaped distributions of parent variables, described in Section 5.5, were obtained in Gajek and Rychlik [32] and Rychlik [84], respectively. We call a symmetric distribution U-shaped if it has nonincreasing and nondecreasing density on the lower and upper halves of its support. Bias of quantile estimation in dependent samples is studied in Section 5.6. Section 5.7 deals with the extreme deviation of expected order statistics under violating the independence assumption.

It is worth pointing out that Papadatos [71] established sharp uniform bounds on distribution functions of order statistics under the assumption that the maxima of the subsamples of a given size have a specified common distribution. The special cases were the independent and arbitrarily dependent identically distributed samples discussed in this book. Balakrishnan et al. [14] presented the best possible mean-variance bounds on the expectations of order statistics from the samples taken without replacement from finite populations. As the size of a population increases to infinity, the bounds reduce to those of the i.i.d. samples from arbitrary populations (cf. Section 4.1). On the other hand, in the case of exhaustive drawing without replacement, these results coincide with ones for arbitrarily dependent samples with arbitrary common marginal distributions, presented in Section 5.2, which are actually attained by exhaustive sampling models. For some quantile bounds on order statistics of dependent observations and respective L-statistics, we refer the reader to Rychlik [84, Section 5].

5.1 Dependent Observations with Given Marginal Distribution

Suppose that arbitrarily dependent $Y_1, \ldots, Y_n$ have a fixed common distribution function F with a finite mean μ_F. We aim to justify bound (2.25) for the expectation of an arbitrary combination of order statistics under all possible interdependencies of observations and to specify the conditions of its attainability. For the formal proof we refer the reader to Rychlik [79] (see also Rychlik [84]).

First we characterize vectors $(\tilde{G}_{1:n}, \ldots, \tilde{G}_{n:n})$ of all possible distribution functions of order statistics $Y_{1:n}, \ldots, Y_{n:n}$ by relations

$$\sum_{j=1}^{n} \tilde{G}_{j:n} = nF, \tag{5.1}$$

$$\tilde{G}_{1:n} \geq \tilde{G}_{2:n} \geq \ldots \geq \tilde{G}_{n:n}. \tag{5.2}$$

5.1 Dependent Observations with Given Marginal Distribution   97

The former immediately follows from

$$\sum_{j=1}^n \mathbf{1}_{(-\infty,x]}(Y_{j:n}) = \sum_{j=1}^n \mathbf{1}_{(-\infty,x]}(Y_j), \quad x \in \Re, \tag{5.3}$$

by taking expectations of both sides of (5.3). The latter is a consequence of relations $Y_{1:n} \leq Y_{2:n} \leq \ldots \leq Y_{n:n}$. There are many ways of constructing ordered variables $Y_{j:n}$, $1 \leq j \leq n$, with distributions satisfying (5.1) and (5.2) (the simplest one consists in taking $\tilde{G}_{j:n}^{-1}(X)$, $1 \leq j \leq n$, for X being a standard uniform variable). There are also many ways of constructing random variables Y_j, $1 \leq j \leq n$, with identical marginal distribution function F, whose order statistics have given distribution functions satisfying (5.1) and (5.2) (the simplest one consists in random rearranging $Y_{j:n}$, $1 \leq j \leq n$).

Now we solve the problem of minimizing $\sum_{j=1}^n c_j \tilde{G}_{j:n}(x)$ for fixed coefficients $\mathbf{c} = (c_1, \ldots, c_n)$ of the L-statistic under study and distribution functions satisfying (5.1) and (5.2) valued at arbitrary point x. This is a linear programming problem that has the solution

$$\min \sum_{j=1}^n c_j \tilde{G}_{j:n}(x) = G_{\mathbf{c}} F(x) \tag{5.4}$$

for $G_{\mathbf{c}}$ being the greatest convex function on the unit interval satisfying (2.26). Recalling Lemma 1, we obtain

$$\begin{aligned}
E_F \sum_{j=1}^n c_j Y_{j:n} &= \int_{-\infty}^{+\infty} y \left(\sum_{j=1}^n c_j \tilde{G}_{j:n}(dy) \right) \\
&\leq \int_{-\infty}^{+\infty} y\, G_{\mathbf{c}} F(dy) \\
&= \int_0^1 F^{-1}(x)\, G_{\mathbf{c}}(dx) \\
&= \int_0^1 F^{-1}(x) g_{\mathbf{c}}(x)\, dx \\
&= (F^{-1}, g_{\mathbf{c}}), \tag{5.5}
\end{aligned}$$

which is the desired conclusion.

Analyzing values $\tilde{G}_{j:n}(x)$, $1 \leq j \leq n$, $x \in \Re$, for the solutions to (5.4), we are able to determine supports and some mutual relations of respective order statistics. They depend on properties of function $G_{\mathbf{c}}$, whose graph is a (broken) line with breaks (if any) at some multiplicities of $1/n$. Let $0 = k_0 < k_1 < \ldots < k_\kappa = n$, $1 \leq \kappa \leq n$, be the sequence of integers such that each k_i/n, $1 \leq i \leq \kappa - 1$, is a breakpoint of $G_{\mathbf{c}}$. Let $0 = l_0 < l_1 <$

$\ldots < l_\lambda = n$, $\kappa \leq \lambda \leq n$, be the sequence of integers for which

$$G_c(j/n) = \sum_{k=1}^{n} c_k$$

holds. We have

$$\{0, n\} \subset \mathcal{K} = \{k_i : 1 \leq i \leq \kappa\}$$
$$\subset \mathcal{L} = \{l_i : 1 \leq i \leq \lambda\}$$
$$\subset \{0, \ldots, n\}.$$

Then equality in (5.4) for all $x \in \Re$ and some $\tilde{G}_{j:n}$ satisfying (5.1) and (5.2) implies

$$P(F^{-1}(k_{i-1}/n) \leq Y_{l_{j-1}+1:n} = Y_{l_j:n} \leq F^{-1}(k_i/n)) = 1 \qquad (5.6)$$

for all $1 \leq i \leq \kappa$, and $1 \leq j \leq \lambda$ such that $k_{i-1} \leq l_{j-1} < l_j \leq k_i$. Relation $\mathcal{L} = \mathcal{K}$ uniquely determines the distributions of all order statistics that solve (5.4) and attain equality in (5.5). Then

$$\tilde{G}_{j:n}(x) = \frac{nF(x) - k_{i-1}}{k_i - k_{i-1}}, \qquad F^{-1}\left(\frac{k_{i-1}}{n}\right) \leq x < F^{-1}\left(\frac{k_i}{n}\right),$$

for $k_{i-1} < j \leq k_i$. If $\mathcal{L} \neq \mathcal{K}$, there are also other solutions. For instance, for the sample mean we have

$$G_c(x) = x,$$
$$\mathcal{K} = \{0, n\},$$
$$\mathcal{L} = \{0, \ldots, n\},$$

and (5.6) simply means that each $Y_{j:n}$ should belong to the domain of Y_1. Indeed,

$$E_F \frac{1}{n} \sum_{j=1}^{n} Y_{j:n} = \int_0^1 F^{-1}(x)\, dx = \mu_F$$

for any type of dependence in variables.

In the special case of single order statistics $Y_{j:n}$, functions G_c and g_c have the forms

$$G_{j:n}(x) = \frac{n}{n+1-j}\left(x - \frac{j-1}{n}\right)_+,$$
$$g_{j:n}(x) = \frac{n}{n+1-j}\mathbf{1}_{[(j-1)/n,1)}(x),$$

respectively, which implies (2.27). Since here

$$\mathcal{K} = \{0, j-1, n\} \subset \mathcal{L} = \{0, 1, \ldots, j-1, n\},$$

5.1 Dependent Observations with Given Marginal Distribution

relation
$$P(Y_{j-1:n} \leq F^{-1}((j-1)/n) \leq Y_{j:n} = Y_{n:n}) = 1 \qquad (5.7)$$
combined with (5.1) and (5.2) are respective conditions for equality. The stochastically largest (i.e., uniformly smallest) distribution function
$$G_{j:n}F(x) = \frac{[nF(x)+1-j]_+}{n+1-j}$$
of the jth order statistic is uniquely determined. There are various ways of constructing dependent samples with the same marginal distribution and the extreme distribution of $Y_{j:n}$. The simplest one is a random rearrangement of

$$\begin{aligned}
Y_{1:n} = \ldots = Y_{j-1:n} &= F^{-1}\left(\frac{j-1}{n}X\right) \\
&\leq F^{-1}\left(\frac{j-1}{n}\right) \\
&\leq F^{-1}\left(\frac{j-1}{n} + \left[1 - \frac{j-1}{n}\right]X\right) \\
&= Y_{j:n} = \ldots = Y_{n:n}
\end{aligned}$$

for some X uniformly distributed on $[0,1]$. For the sample minimum, (2.27) and (5.7) imply the trivial claims $E_F Y_{1:n} \leq \mu_F$, becoming the equality for identical observations $Y_{1:n} = Y_{n:n} = Y_1$. Condition (5.7) excludes the possibility of constructing an absolutely continuous joint distribution of the sample with the stochastically maximal jth order statistic except for the sample maximum.

In the first paper in this field of research Mallows [53] constructed a density function of the sample with identical uniform marginals that provide the maximal expectation of the sample maximum. Lai and Robbins [47] extended the construction to arbitrary possibly nonidentical marginal distributions. Lai and Robbins [48] and Tchen [98] constructed infinite sequences of variables with identical and arbitrary distributions, respectively, such that all sample maxima are stochastically maximal. Bounds (2.27) for general order statistics of identically distributed samples were proved independently in Caraux and Gascuel [19] and Rychlik [76]. In the former, some inequalities for nonidentically distributed observations were presented (with conditions for equality established in Rychlik [82]). In the latter, the problem of constructing sequences with stochastically extreme order statistics was also discussed. Asymptotic properties of sequences of stochastically extreme maxima and other order statistics were studied in Lai and Robbins [48] and Rychlik [77], respectively.

100   5. Order Statistics of Dependent Observations

In the rest of this chapter we present tight bounds for expected order and L-statistics of dependent samples for various families of marginal distributions. We also indicate the marginals for which the bounds are attained. Formally, in each case one should say that these are attained by the joint distributions specified for arbitrary marginals and fixed L-statistics by (5.1) and (5.2) with (5.6) (replaced by (5.7) for single order statistics in particular), and the given marginal. However, we drop repeating the reference to the construction of the joint probability and confine ourselves to describing the optimal marginal.

5.2   General and Symmetric Distributions

We first study the optimal bounds for general L-statistics based on dependent samples with arbitrary common marginal distribution F. For an arbitrarily fixed sequence of coefficients $\mathbf{c} = (c_1, \ldots, c_n)$, we construct function $g_\mathbf{c}$ that defines the functional of the expectation of the respective L-statistic (see (2.25) and (5.5)) by differentiating function $G_\mathbf{c}$ determined by (2.26). Using the notation of Section 5.1, we have

$$g_\mathbf{c}(x) = \sum_{j=1}^{n} d_j \mathbf{1}_{[(j-1)/n, j/n)}(x) = \sum_{i=1}^{\kappa} d_{k_i} \mathbf{1}_{[k_{i-1}/n, k_i/n)}(x) \qquad (5.8)$$

for a nondecreasing sequence d_j, $1 \leq j \leq n$, with the specified increasing subsequence d_{k_i}, $1 \leq i \leq \kappa \leq n$, of distinct values. Precisely, we have

$$d_j = d_{k_i} = \frac{n\left[G_\mathbf{c}\left(\frac{k_i}{n}\right) - G_\mathbf{c}\left(\frac{k_{i-1}}{n}\right)\right]}{k_i - k_{i-1}} = \frac{n}{k_i - k_{i-1}} \sum_{k=k_{i-1}+1}^{k_i} c_k, \qquad (5.9)$$

$k_{i-1} + 1 \leq j \leq k_i$, $1 \leq i \leq \kappa$. Also, put

$$\bar{d} = \frac{1}{n}\sum_{j=1}^{n} d_j = \sum_{j=1}^{n} c_j = G_\mathbf{c}(1) = \int_0^1 g_\mathbf{c}(x)\, dx. \qquad (5.10)$$

We claim that the optimal mean-variance bound for the expectation of a given L-statistic amounts to the Euclidean norm of the respective vector $(d_j - \bar{d})/n$, $1 \leq j \leq n$.

Theorem 19 (general distributions) *Under notation (5.8) to (5.10), with $G_\mathbf{c}$ defined in (2.26), we have*

$$E_F \sum_{j=1}^{n} c_j \frac{Y_{j:n} - \mu_F}{\sigma_F} \leq C = C^0(\mathbf{c}) = \left[\frac{1}{n}\sum_{j=1}^{n} d_j^2 - \bar{d}^2\right]^{1/2}. \qquad (5.11)$$

If $C = 0$, then (5.11) becomes the equality for degenerate distributions. Otherwise the equality holds for the κ-point marginal distribution

$$P\left(Y_1 = \mu + \sigma \frac{d_{k_i} - \bar{d}}{C}\right) = \frac{k_i - k_{i-1}}{n}, \quad 1 \leq i \leq \kappa. \quad (5.12)$$

PROOF. Verification of (5.11)

$$\begin{aligned}
E_F \sum_{j=1}^{n} c_j(Y_{j:n} - \mu) &= \int_0^1 F^{-1}(x)[g_c(x) - \bar{d}] \, dx \\
&= \int_0^1 [F^{-1}(x) - \mu][g_c(x) - \bar{d}] \, dx \\
&\leq \left\{\int_0^1 [F^{-1}(x) - \mu]^2 \, dx \int_0^1 [g_c(x) - \bar{d}]^2 \, dx\right\}^{1/2} \\
&= C\sigma
\end{aligned}$$

is based on (2.25), (5.10), (2.10), and (5.8). The Schwarz inequality provides the sharp bound here, because $g_c - \bar{d}$ is nondecreasing, and so

$$F^{-1}(x) - \mu = \alpha(g_c(x) - \bar{d})$$

with $\alpha = \sigma/C > 0$ actually defines a quantile function with the desired mean and variance. An elementary algebra enables us to derive the respective distribution (5.12). ■

Especially, for the single order statistics we have

Corollary 1 (general distributions) *Inequality*

$$\frac{E_F Y_{j:n} - \mu_F}{\sigma_F} \leq C = C^0(j,n) = \left(\frac{j-1}{n+1-j}\right)^{1/2} \quad (5.13)$$

is sharp and becomes the equality for the two-point marginal

$$P\left(Y_1 = \mu - \frac{\sigma}{C}\right) = \frac{j-1}{n} = 1 - P(Y_1 = \mu + \sigma C). \quad (5.14)$$

Bounds (5.13) for the sample maximum and general j were presented in Arnold [2] and Gascuel and Caraux [34], respectively. We proceed now to present analogous results for symmetrically distributed random variables. Below in Theorem 20 we omit the formal presentation of the distribution that attains the bound, because this needs introducing a rather complicated notation.

Theorem 20 (symmetric distributions) *Inequality*

$$E_F \sum_{j=1}^n c_j \frac{Y_{j:n} - \mu_F}{\sigma_F} \leq C^s(\mathbf{c}) = \left[\frac{1}{2n} \sum_{j=\lfloor(n+3)/2\rfloor}^n (d_j - d_{n+1-j})^2\right]^{1/2}, \quad (5.15)$$

where $\lfloor \cdot \rfloor$ denotes the floor of a number, is tight. Equality is attained by a unique (up to location-scale transformations) marginal symmetric distribution supported on n points at most whose probabilities are multiplicities of $1/n$.

PROOF. Inequality (5.15) follows from

$$E_F \sum_{j=1}^n c_j (Y_{j:n} - \mu) = \int_{1/2}^1 [F^{-1}(x) - \mu] g_c^s(x)\, dx \quad (5.16)$$

with

$$g_c^s(x) = \frac{g_c(x) - g_c(1 - x-)}{2}$$

$$= \sum_{j=\lfloor(n+3)/2\rfloor}^n \frac{d_j - d_{n+1-j}}{2} \mathbf{1}_{[(j-1)/n, j/n)}(x), \quad (5.17)$$

$1/2 \leq x < 1$, and the Schwarz inequality. Note that (5.17) is a nonnegative nondecreasing piecewise constant function with $\lfloor(n+1)/2\rfloor$ values at most. Thus its antisymmetric extension onto $[0, 1/2)$ is also a nondecreasing step function with n values at most and jumps at some points j/n, $1 \leq j \leq n-1$. The extension coincides with the quantile function of the extreme distribution providing equality in (5.15) up to an affine transformation. This justifies the latter claim of Theorem 20. ■

Although the algorithm of determining g_c and g_c^s is simple, we are not able to write explicitly respective formulae for general linear combinations of order statistics. For the single order statistics, however, we obtain

$$g_{j:n}^s(x) = \frac{n}{n + 1 - j} \mathbf{1}_{[\max\{(j-1)/n, 1-(j-1)/n\}, 1)}(x), \quad 1/2 \leq x < 1. \quad (5.18)$$

By an easy computation, we conclude

Corollary 2 (symmetric distributions) *Inequality*

$$\frac{E_F Y_{j:n} - \mu_F}{\sigma_F} \leq C = \left[\frac{n}{2(n+1-j)} \min\left\{\frac{j-1}{n+1-j}, 1\right\}\right]^{1/2} \quad (5.19)$$

is sharp and becomes the equality for the three-point marginal distribution

$$P(Y_1 = \mu) = 2\left|\frac{j-1}{n} - \frac{1}{2}\right|, \quad (5.20)$$

$$P(Y_1 = \mu \pm \sigma C) = \min\left\{\frac{j-1}{n}, 1 - \frac{j-1}{n}\right\}. \quad (5.21)$$

Inequality (5.19) and its special case for $j = n$ can be found in Gascuel and Caraux [34] and Arnold [2], respectively. Here we merely mention the second moment bounds for general L-statistics

$$\begin{aligned} E_F \sum_{j=1}^{n} c_j Y_{j:n} &\leq \int_0^1 F^{-1}(x)[g_c(x)]_+ \, dx \\ &\leq \|(g_c)_+\| m_F \\ &= \left[\frac{1}{n} \sum_{j=1}^{n} (d_j)_+^2 \right]^{1/2} m_F \end{aligned} \qquad (5.22)$$

and for single order statistics

$$E_F Y_{j:n} \leq \left(\frac{n}{n+1-j} \right)^{1/2} m_F \qquad (5.23)$$

of nonnegative samples. Bounds (5.22) and (5.23) are attained by

$$P(Y_1 = 0) = \frac{k_{i_0}}{n}, \qquad (5.24)$$

$$P\left(Y_1 = m_F \frac{d_{ki}}{\|(g_c)_+\|} \right) = \frac{k_i - k_{i-1}}{n}, \quad i_0 < i \leq \kappa, \qquad (5.25)$$

with $i_0 = \max\{0 \leq i \leq \kappa : d_{k_i} \leq 0\}$, and

$$P(Y_1 = 0) = \frac{j-1}{n} = 1 - P\left(Y_1 = \left[\frac{n}{n+1-j} \right]^{1/2} m_F \right), \qquad (5.26)$$

respectively. If $d_n \leq 0$ and so $\|(g_c)_+\| = 0$, then (5.24) and (5.25) are clearly replaced by the Dirac measure at 0. In fact, (5.24) with (5.25), and (5.26) are not unique solutions. The second relation in (5.22) shows that F^{-1} may have various forms on $[0, k_{i_0}/n)$ provided that nonnegativity, nondecrease, and moment conditions are not violated (see Rychlik [81] for more details).

Rychlik [81] (and Arnold [3] for the cases of sample maximum $Y_{n:n}$ and range $Y_{n:n} - Y_{1:n}$) presented more general sharp bounds in terms of central absolute moments of various orders based on the Hölder inequality instead of the Schwarz one. These are also attainable by discrete marginal distributions with probabilities k_i/n for some integer k_i. This form of solution allows us to deduce analogous inequalities for deterministic sequences. Randomly rearranging a sequence of (not necessarily distinct) numbers $y_1, \ldots, y_n$, we obtain a random sequence of dependent identically distributed random variables with expectation, second raw moment, variance

$$\bar{y} = \frac{1}{n} \sum_{j=1}^{n} y_j,$$

$$m^2 = \frac{1}{n}\sum_{j=1}^{n} y_j^2,$$

$$s^2 = \frac{1}{n}\sum_{j=1}^{n}(y_j - \bar{y})^2 = m^2 - \bar{y}^2,$$

respectively, and deterministic order statistics $y_{1:n} \leq \ldots \leq y_{n:n}$. Using (5.11), (5.15), and (5.22), we conclude optimal bounds

$$\sum_{j=1}^{n}\frac{c_j(y_{j:n} - \bar{y})}{s} \leq \left[\frac{1}{n}\sum_{j=1}^{n} d_j^2 - \bar{d}^2\right]^{1/2}, \tag{5.27}$$

$$\sum_{j=1}^{n}\frac{c_j(y_{j:n} - \bar{y})}{s} \leq \left[\frac{1}{2n}\sum_{j=\lfloor(n+3)/2\rfloor}^{n}(d_j - d_{n+1-j})^2\right]^{1/2}, \tag{5.28}$$

$$\sum_{j=1}^{n}\frac{c_j y_{j:n}}{m} \leq \left[\frac{1}{n}\sum_{j=1}^{n}(d_j)_+^2\right]^{1/2}, \tag{5.29}$$

for general, symmetric, and nonnegative sequences of numbers, respectively. Sharpness of (5.27) is verified by putting $y_j = d_j$, $1 \leq j \leq n$, and with reference to (5.12).

The problem of best deterministic bounds for L-statistics in terms of various sample parameters has a long history. We confine ourselves to those derived by means of the Schwarz inequality. Samuelson [92] raised and solved the problem of how much a single observation can deviate from the sample mean in the standard deviation units. Samuelson's paper stimulated intensive investigations of the problem and its modifications: alternative proofs, rediscoveries of earlier results, and extensions. Six different proofs were reviewed by Arnold and Balakrishnan [5]. The earliest proofs found in the literature were due to Thompson [99] and Scott [94]. We do not attempt to present a complete record of consecutive contributions, referring the reader to Arnold [4] for a comprehensive bibliography, and Olkin [66] for a recent review, with yet another proof. Scott [94] established the bound for deviations of $y_{n-1:n}$ from the mean in the standard deviation units. The bounds for arbitrary order statistics follow directly from Mallows and Richter [54], and were explicitly stated by Boyd [18] and Hawkins [39]. Mallows and Richter [54] established the inequalities for selection differentials $\sum_{i=1}^{j} y_{i:n}/j$, and $\sum_{i=n+1-k}^{n} y_{i:n}/k$, and their differences. The respective results for $y_{n:n} - y_{1:n}$, $y_{n-1:n} - y_{1:n}$, and the differences of arbitrary order statistics were derived by Nair [63], David et al. [24], and Fahmy and Proschan [27] (implicitly in Arnold and Groeneveld [6]), respectively, and for the L-statistics with nondecreasing coefficients by David [23]. Bounds (5.27) through (5.29) for general L-statistics come from Rychlik [81].

5.3 Distributions with Monotone Density and Failure Rate

Determining mean-variance bounds for $F \succeq_c W$ by the projection method, we look for the element of $\mathcal{C}^{\nearrow}_{\succeq_c W}$ least distant from the function

$$h(x) = \frac{n}{n+1-j} \mathbf{1}_{[(j-1)/n, 1)} W(x)$$
$$= \frac{n}{n+1-j} \mathbf{1}_{[W^{-1}((j-1)/n), d_W)}(x).$$

Applying Lemmas 4 and 5 with $b = W^{-1}((j-1)/n)$ and $c = d_W$, we obtain the solution. Note that there are no trivial zero projections and bounds for $b > a_W$, because

$$E_W X < \mu_W(b) = E_W(X|X > b)$$

contradicts (3.8). Hence the projection is linear increasing with a possible constant left part, and the condition for distinguishing the cases is given in the last part of Lemma 5.

Theorem 21 ($F \succeq_c W$) *If for $b = W^{-1}((j-1)/n)$ we have*

$$\mu_W(b) \leq \mu_W + \frac{\sigma_W^2}{\mu_W - a_W}, \tag{5.30}$$

then inequality

$$\frac{E_F Y_{j:n} - \mu_F}{\sigma_F} \leq \frac{\mu_W(b) - \mu_W}{\sigma_W} \tag{5.31}$$

holds true and becomes the equality for

$$F(x) = W\left(\mu_W + \sigma_W \frac{x - \mu}{\sigma}\right). \tag{5.32}$$

Otherwise there exists a unique $\beta_* = \beta_*(j, n) \in (a_W, b)$ satisfying equation

$$\sigma_W^2(\beta) = [\mu_W(b) - \mu_W(\beta)][\mu_W(\beta) - \beta] \tag{5.33}$$

such that

$$\frac{E_F Y_{j:n} - \mu_F}{\sigma_F} \leq C^0_{\succeq_c W}(j, n) = \frac{\hat{\vartheta}_W(\beta_*)}{\hat{\eta}_W(\beta_*)} \tag{5.34}$$

for

$$\hat{\eta}_W(\beta) = E_W(X - \beta)_+ = \int_\beta^d (x - \beta) w(x)\, dx, \tag{5.35}$$

$$\hat{\vartheta}_W^2(\beta) = \text{Var}_W(X - \beta)_+$$
$$= \int_\beta^d (x - \beta)^2 w(x)\, dx - \left[\int_\beta^d (x - \beta) w(x)\, dx\right]^2 \tag{5.36}$$

(cf. (4.127) and (4.129)). Bound (5.34) is attained by

$$F(x) = W\left(\beta_* + \hat{\eta}_W(\beta_*) + \hat{\vartheta}_W(\beta_*)\frac{x-\mu}{\sigma}\right) \mathbf{1}_{[\mu - \sigma \hat{\eta}_W(\beta_*)/\hat{\vartheta}_W(\beta_*),\infty)}(x). \tag{5.37}$$

Note that $\sup \mathrm{E}_W Y_{j:n} = \mu_W(b)$ which confirms the first statement of Theorem 21. Condition (5.30) is certainly satisfied by small order statistics. Definitions (5.35) and (5.36) are analogues of (3.59) and (3.60), respectively, valued at (β, β). Distribution function (5.37) has a jump of height $W(\beta) < (j-1)/n$ at $\mu - \sigma \hat{\eta}/\hat{\vartheta}$ and a density right to the point.

Proposition 14 (decreasing density) *If* $(j-1)/n \leq 1/3$, *then*

$$\frac{\mathrm{E}_F Y_{j:n} - \mu_F}{\sigma_F} \leq \sqrt{3}\frac{j-1}{n}, \tag{5.38}$$

which becomes the equality for F being the uniform distribution function on $[\mu - \sqrt{3}\sigma, \mu + \sqrt{3}\sigma]$.

Otherwise

$$\frac{\mathrm{E}_F Y_{j:n} - \mu_F}{\sigma_F} \leq \frac{1}{3}\left(\frac{9j-9-n}{n+1-j}\right)^{1/2}, \tag{5.39}$$

and the equality holds for the mixture of the Dirac distribution at $\mu - 3\sigma [(n+1-j)/(9j-9-n)]^{1/2}$ and the uniform distribution on

$$\left[\mu - 3\sigma\left(\frac{n+1-j}{9j-9-n}\right)^{1/2}, \mu + \sigma\frac{3j \; 3 \mid n}{[(n+1-j)(9j-9-n)]^{1/2}}\right]$$

with respective probabilities $3(j-1)/(2n) - 1/2$ and $3(n+1-j)/(2n)$.

Proposition 15 (decreasing failure rate) *If* $(j-1)/n \leq 1 - e^{-1} \approx 0.63212$, *then*

$$\frac{\mathrm{E}_F Y_{j:n} - \mu_F}{\sigma_F} \leq \ln\frac{n}{n+1-j}, \tag{5.40}$$

and the equality holds in (5.40) for the exponential distribution with location $\mu - \sigma$ and scale σ.

Otherwise for

$$\gamma = \gamma_V(j,n) = \left(1 - \frac{j-1}{n}\right)e,$$

we have

$$\frac{\mathrm{E}_F Y_{j:n} - \mu_F}{\sigma_F} \leq C = C^0_{\geq_c V}(j,n) = \left[\frac{2}{\gamma_V(j,n)} - 1\right]^{1/2}, \tag{5.41}$$

which becomes the equality for

$$F(x) = 1 - \gamma\exp\left(-\gamma - \gamma C\frac{x-\mu}{\sigma}\right)\mathbf{1}_{[\mu-\sigma/C,\infty)}(x). \tag{5.42}$$

5.3 Distributions with Monotone Density and Failure Rate

Note that (5.42) is the mixture of the exponential distribution with location $\mu - \sigma/C$ and scale $\sigma/(\gamma C)$, with probability γ, and the Dirac measure at $\mu - \sigma/C$ with probability $1 - \gamma$. This is a DFR life distribution if $\mu = \sigma/C$. For small order statistics (5.40) is attained by a life exponential distribution with $\mu = \sigma$.

Now we consider analogous bounds for $F \preceq_c W$. First we try to describe $P^{\nearrow}_{\preceq_c W} h$ for

$$h(x) = \frac{n}{n+1-j} \mathbf{1}_{[W^{-1}((j-1)/n), d_W)}(x)$$
$$= \frac{1}{1-W(b)} \mathbf{1}_{[b,d)}(x).$$

Results of Lemmas 6 and 7 may be applied here. However, setting $c = d_W$ also results in reduction of the number of parameters of functions that are the candidates for being projections. In contrast to the conclusion of Lemma 6, the level at which the linear functions are possibly broken is specified and coincides with the level of the original h.

Lemma 22 *If $h(x) = M\mathbf{1}_{[b,d)}(x)$, then for every $g \in C^{\nearrow}_{\preceq_c W}$ there exists $g_{\alpha\beta} \in C^{\nearrow}_{\preceq_c W}$ defined as*

$$g_{\alpha\beta}(x) = M + \alpha \min\{x - \beta, 0\}$$

for some $\alpha \geq 0$ and $\beta \geq b$ such that

$$\|g_{\alpha\beta} - h\| \leq \|g - h\|.$$

Lemma 23 *Under the assumptions of Lemma 22 and (3.7), we have*

$$P^{\nearrow}_{\preceq_c W} h(x) = \frac{1}{\int_b^d w(x)\, dx} + \alpha_* \min\{x - \beta_*, 0\}, \qquad (5.43)$$

where $\beta_* > b$ is the unique solution to the equation

$$\beta \left[\int_a^b w(x)\, dx \int_a^{\bar{\beta}} xw(x)\, dx - \int_a^b xw(x)\, dx \int_a^{\bar{\beta}} w(x)\, dx \right]$$
$$= \int_a^b w(x)\, dx \int_a^{\bar{\beta}} x^2 w(x)\, dx - \int_a^b xw(x)\, dx \int_a^{\bar{\beta}} xw(x)\, dx, \qquad (5.44)$$

with $\bar{\beta} = \min\{\beta, d\}$, and

$$\alpha_* = \alpha_*(\beta_*) = \frac{\int_a^b w(x)\, dx}{\int_b^d w(x)\, dx \int_a^{\min\{\beta_*, d\}} (\beta_* - x)w(x)\, dx}. \qquad (5.45)$$

5. Order Statistics of Dependent Observations

Precisely, if

$$d \le \check{\beta} = \frac{\int_a^b w(x)\,dx \int_a^d x^2 w(x)\,dx - \int_a^b xw(x)\,dx \int_a^d xw(x)\,dx}{\int_a^b w(x)\,dx \int_a^d xw(x)\,dx - \int_a^b xw(x)\,dx}, \quad (5.46)$$

(i.e., the left-hand side of (5.44) is not greater than the right-hand side for $\beta = d$), then $\beta_* = \check{\beta}$, and (5.43) is linear. This may happen if d is finite and b is close to d. A probabilistic interpretation of Lemma 23 is given in Theorem 22. For $a < \beta < d$, we define

$$\check{\mu}_W(\beta) = E_W(X|X < \beta) = \frac{\int_a^\beta xw(x)\,dx}{\int_a^\beta w(x)\,dx}, \quad (5.47)$$

$$\check{\sigma}_W^2(\beta) = \text{Var}_W(X|X < \beta)$$

$$= \frac{\int_a^\beta x^2 w(x)\,dx - \check{\mu}_W^2(\beta)}{\int_a^\beta w(x)\,dx}, \quad (5.48)$$

$$\check{\eta} = \check{\eta}_W(\beta) = E_W(\beta - X)_+ = \int_a^\beta (\beta - x)w(x)\,dx, \quad (5.49)$$

$$\check{\vartheta}^2 = \check{\vartheta}_W^2(\beta) = \text{Var}_W(\beta - X)_+$$

$$= \int_a^\beta (\beta - x)^2 w(x)\,dx - \left[\int_a^\beta (\beta - x)w(x)\,dx\right]^2 \quad (5.50)$$

(cf. (3.20) and (3.21) as well as (5.35) and (5.36)).

Theorem 22 ($F \preceq_c W$) *If for $b = W^{-1}((j-1)/n)$*

$$\frac{m_W^2 - \check{\mu}_W(b)\mu_W}{\mu_W - \check{\mu}_W(b)} < d \quad (5.51)$$

holds, then there exists a unique $\beta_* \in (b, d)$ satisfying the equation

$$\check{\sigma}_W^2(\beta) = [\check{\mu}_W(\beta) - \check{\mu}_W(b)][\beta - \check{\mu}_W(\beta)] \quad (5.52)$$

such that

$$\frac{E_F Y_{j:n} - \mu_F}{\sigma_F} \le C = C_{\preceq_c W}^0(j, n) = \frac{j-1}{n+1-j} \frac{\check{\vartheta}_W(\beta_*)}{\check{\eta}_W(\beta_*)} \quad (5.53)$$

(cf. (5.33) and (5.34)). The equality in (5.53) holds if

$$F(x) = \begin{cases} W\left(\beta_* - \check{\eta}(\beta_*) + \check{\vartheta}(\beta_*)\frac{x-\mu}{\sigma}\right) & \text{if } \frac{x-\mu}{\sigma} < \frac{\check{\eta}(\beta_*)}{\check{\vartheta}(\beta_*)}, \\ 1, & \text{if } \frac{x-\mu}{\sigma} \ge \frac{\check{\eta}(\beta_*)}{\check{\vartheta}(\beta_*)}. \end{cases} \quad (5.54)$$

Otherwise we have (5.31) which becomes the equality for (5.32).

5.3 Distributions with Monotone Density and Failure Rate

Distribution function (5.54) jumps from $W(\beta_*) > (j-1)/n$ to 1 at its right endpoint. If $W = U$, both cases are possible. Since $d_V = +\infty$, the first one is only applicable for the IFR distributions.

Proposition 16 (increasing density) *If $(j-1)/n < 2/3$, then*

$$\frac{E_F Y_{j:n} - \mu_F}{\sigma_F} \leq \frac{[(j-1)(8n - 9j + 9)]^{1/2}}{3(n+1-j)}, \quad (5.55)$$

which becomes the equality for the combination of the uniform distribution on

$$\left[\mu - \sigma \frac{4n - 3j + 3}{[(j-1)(8n - 9j + 9)]^{1/2}}, \mu + \sigma \left[\frac{9j - 9}{8n - 9j + 9}\right]^{1/2}\right],$$

and an atom at the right end of the interval, with respective coefficients $3(j-1)/(2n)$ and $1 - 3(j-1)/(2n)$.

If $(j-1)/n \geq 2/3$, then (5.38) holds with the equality for the uniform distribution on $[\mu - \sqrt{3}\sigma, \mu + \sqrt{3}\sigma]$.

Proposition 17 (increasing failure rate) *For $\beta_* > \ln[n/(n+1-j)]$ defined by the equation*

$$\frac{1 - (1+\beta)e^{-\beta}}{e^{-\beta} - 1 + \beta} = \frac{n+1-j}{j-1} \ln \frac{n}{n+1-j} \quad (5.56)$$

and

$$\theta^2 = \theta_V^2(\beta_*) = 1 - 2\beta_* e^{-\beta_*} - e^{-2\beta_*} \quad (5.57)$$

we have

$$\frac{E_F Y_{j:n} - \mu_F}{\sigma_F} \leq \frac{j-1}{n+1-j} \frac{\theta_V(\beta_*)}{e^{-\beta_*} - 1 + \beta_*}. \quad (5.58)$$

The equality in (5.58) holds if

$$F(x) = \begin{cases} 0, & \text{if } \frac{x-\mu}{\sigma} \leq -\frac{1-e^{-\beta_*}}{\theta}, \\ 1 - \exp(-1 + e^{-\beta_*} - \theta \frac{x-\mu}{\sigma}), & \text{if } -\frac{1-e^{-\beta_*}}{\theta} \leq \frac{x-\mu}{\sigma} < \frac{e^{-\beta_*} - 1 + \beta_*}{\theta}, \\ 1, & \text{if } \frac{x-\mu}{\sigma} \geq \frac{e^{-\beta_*} - 1 + \beta_*}{\theta}. \end{cases} \quad (5.59)$$

This is the exponential distribution with location $\mu - \sigma(1 - e^{-\beta_*})/\theta$ and scale σ/θ, right truncated at $\mu + \sigma(e^{-\beta_*} - 1 + \beta_*)/\theta$. The jump probability is $e^{-\beta_*}$.

Likewise, for determining respective second moment bounds for life distributions $F \succeq_c W$ and $F \preceq_c W$, we need to find best approximations of

$$h(x) = \frac{n}{n+1-j} \mathbf{1}_{[(j-1)/n, 1)} W(x)$$

by nondecreasing convex and concave functions that additionally obey the condition of vanishing at a_W. We assume here that W is a life distribution and $a_W = 0$. The respective projections have similar parametric forms

$$P^+_{\succeq_c W} h(x) = \alpha(x - \beta)_+,$$

$$P^+_{\succeq_c W} h(x) = \frac{n}{n+1-j} \min\{\frac{x}{\alpha}, 1\}.$$

In the first case, under the conditions of the latter statement of Theorem 21, the optimal parameter β defined in (5.33) is positive which implies that $P^\nearrow_{\succeq_c W} h \in C^+_{\succeq_c W}$ provides the solution to our problem. Then we have

$$\|P^+_{\succeq_c W} h\|^2 = \|P^0_{\succeq_c W} h\|^2 + 1 = \frac{\int_\beta^d (x - \beta)^2 w(x)\, dx}{\left[\int_\beta^d (x - \beta) w(x)\, dx\right]^2} \qquad (5.60)$$

(cf. (5.34) to (5.36)). Otherwise $P^+_{\succeq_c W} h$ is the linear function crossing the origin, with the optimal slope and norm

$$\alpha_*(0) = \frac{\int_b^d x w(x)\, dx}{\int_b^d w(x)\, dx \int_0^d x^2 w(x)\, dx} = \frac{\mu_W(b)}{m_W^2}, \qquad (5.61)$$

$$\|P^+_{\succeq_c W} h\| = \frac{\int_b^d x w(x)\, dx}{\int_b^d w(x)\, dx \left[\int_0^d x^2 w(x)\, dx\right]^{1/2}} = \frac{\mu_W(b)}{m_W}, \qquad (5.62)$$

respectively.

Theorem 23 ($F \succeq_c W$) *If for $b = W^{-1}((j-1)/n)$*

$$\mu_W(b) \leq \frac{m_W^2}{\mu_W} \qquad (5.63)$$

holds true (cf. (5.30)), then

$$\frac{E_F Y_{j:n}}{m_F} \leq \frac{\mu_W(b)}{m_W}, \qquad (5.64)$$

(cf. (5.62)), which is attained for

$$F(x) = W\left(m_W \frac{x}{m_F}\right).$$

Otherwise for $\beta_* \in (0, b)$ determined by (5.33) with

$$\hat{\nu}^2 = \hat{\nu}_W^2(\beta_*) = E_W[(X - \beta_*)_+]^2 = \int_{\beta_*}^d (x - \beta_*)^2 w(x)\, dx \qquad (5.65)$$

we have
$$\frac{E_F Y_{j:n}}{m_F} \leq \frac{\hat{\nu}_W(\beta_*)}{\mu_W(\beta_*) - \beta_*}, \tag{5.66}$$

which becomes the equality for
$$F(x) = W\left(\beta_* + \hat{\nu}\frac{x}{m}\right)\mathbf{1}_{[0,\infty)}(x). \tag{5.67}$$

Again, we point out analogies between Theorems 21 and 23. Under equivalent conditions (5.30) and (5.63), we obtain analogous bounds (5.31) and (5.64), respectively. In the opposite case, the bounds are determined by the same parameter β_* and related by (5.60). Moreover, both the extreme distributions (5.37) and (5.67) are location-scale modifications of W with identical jump $W(\beta_*)$ at the left support end.

Proposition 18 (decreasing density) *If $(j-1)/n \leq 1/3$, then*
$$\frac{E_F Y_{j:n}}{m_F} \leq \frac{\sqrt{3}}{2}\left(1 + \frac{j-1}{n}\right), \tag{5.68}$$

where the equality holds if F is the uniform distribution function on the interval $[0, \sqrt{3}m_F]$.

Otherwise
$$\frac{E_F Y_{j:n}}{m_F} \leq \frac{2}{3}\left(\frac{2n}{n+1-j}\right)^{1/2}, \tag{5.69}$$

which becomes the equality for the mixture of the atom at 0 with probability $3(j-1)/(2n) - 1/2$ and the uniform distribution on the interval $\left[0, m_F\left[2n/(n+1-j)\right]^{1/2}\right]$ with probability $3(n+1-j)/(2n)$.

Proposition 19 (decreasing failure rate) *If $(j-1)/n \leq 1 - e^{-1} \approx 0.63212$, then*
$$\frac{E_F Y_{j:n}}{m_F} \leq \frac{1}{\sqrt{3}}\left(\ln\frac{n}{n+1-j} + 1\right), \tag{5.70}$$

which is the equality for the exponential life distribution with scale $m_F/\sqrt{2}$.

Otherwise
$$\frac{E_F Y_{j:n}}{m_F} \leq \left[\frac{2n}{(n+1-j)e}\right]^{1/2}. \tag{5.71}$$

The bound becomes the equality for the mixture of the exponential distribution with scale $m_F\{n/[2e(n+1-j)]\}^{1/2}$ and zero, with weights $[1-(j-1)/n]e$ and $1 - [1-(j-1)/n]e$, respectively.

Projection of $h = M\mathbf{1}_{[b,d)}$ for $M = n/(n+1-j)$, $0 = a_W < b = W^{-1}((j-1)/n) < d_W$ onto $\mathcal{C}^+_{\preceq_c W}$, desired for calculating second moment bounds for samples with life distributions satisfying $F \preceq_c W$, is described in Lemma 24 (cf. Rychlik [84, Theorem 11(b), p. 131]).

Lemma 24 *Projection of $h = M\mathbf{1}_{[b,d)}$ onto $C^+_{\preceq_c W}$ in $L^2([0, d_W), w(x)\, dx)$ has the form*

$$P^+_{\preceq_c W} h(x) = M \min\{x/\alpha_*, 1\}. \tag{5.72}$$

If

$$d \int_b^d x w(x)\, dx > \int_0^d x^2 w(x)\, dx, \tag{5.73}$$

then $\alpha_* \in (b, d)$ is the unique solution to

$$\alpha \int_b^\alpha x w(x)\, dx = \int_0^\alpha x^2 w(x)\, dx. \tag{5.74}$$

Otherwise

$$\alpha_* = \frac{\int_0^d x^2 w(x)\, dx}{\int_b^d x w(x)\, dx} \geq d. \tag{5.75}$$

Under (5.73) the projection is actually a broken line with a break at α_* that belongs to the domain of h. Then

$$\begin{aligned}
\|P^+_{\preceq_c W} h\|^2 &= M^2 \left[\frac{1}{\alpha_*^2} \int_0^{\alpha_*} x^2 w(x)\, dx + \int_{\alpha_*}^d w(x)\, dx \right] \\
&= M^2 \left[\frac{1}{\alpha_*} \int_b^{\alpha_*} x w(x)\, dx + \int_{\alpha_*}^d w(x)\, dx \right] \quad (5.76) \\
&= M^2 \gamma^2, \tag{5.77}
\end{aligned}$$

say, and

$$\frac{P^+_{\preceq_c W} h(x)}{\|P^+_{\preceq_c W} h\|} = \frac{1}{\gamma} \min\left\{ \frac{x}{\alpha_*}, 1 \right\}. \tag{5.78}$$

If (5.73) is not true (which is possible for $d_W < \infty$ only), then the projection is actually linear and satisfies

$$\|P^+_{\preceq_c W} h\| = M \frac{\int_b^d x w(x)\, dx}{\left[\int_0^d x^2 w(x)\, dx \right]^{1/2}}, \tag{5.79}$$

$$\frac{P^+_{\preceq_c W} h(x)}{\|P^+_{\preceq_c W} h\|} = \frac{x}{\left[\int_0^d x^2 w(x)\, dx \right]^{1/2}}, \tag{5.80}$$

Theorem 24 ($F \preceq_c W$) *If*

$$\frac{n}{n+1-j} \frac{m_W^2}{\mu_W(b)} < d_W \tag{5.81}$$

for $b = W^{-1}((j-1)/n)$, then there exists a unique $b < \alpha_* < d_W$ solving

$$W(\alpha) \check{\mu}_W(\alpha)[\alpha - \check{\mu}_W(\alpha)] = W(\alpha) \check{\sigma}_W^2(\alpha) + [(j-1)/n] \alpha \check{\mu}_W(b) \tag{5.82}$$

5.3 Distributions with Monotone Density and Failure Rate

(cf. (5.47) and (5.48)) such that for

$$\bar{\nu}^2 = \bar{\nu}_W^2(\alpha_*) = E_W(\min\{X, \alpha_*\})^2$$
$$= \int_0^{\alpha_*} x^2 w(x)\,dx + \alpha_*^2 \int_{\alpha_*}^d w(x)\,dx \qquad (5.83)$$

we have

$$\frac{E_F Y_{j:n}}{m_F} \leq \frac{n}{n+1-j} \frac{\bar{\nu}_W(\alpha_*)}{\alpha_*}. \qquad (5.84)$$

The equality in (5.84) holds for

$$F(x) = \begin{cases} W\left(\bar{\nu}\frac{x}{m_F}\right), & \text{if } \frac{x}{m_F} < \frac{\alpha_*}{\bar{\nu}}, \\ 1, & \text{if } \frac{x}{m_F} \geq \frac{\alpha_*}{\bar{\nu}}. \end{cases} \qquad (5.85)$$

If (5.81) does not hold, then bound

$$\frac{E_F Y_{j:n}}{m_F} \leq \frac{\mu_W(b)}{m_W} \qquad (5.86)$$

is attained by $F(x) = W(m_W x/m_F)$.

Relation (5.81) is always true for W supported on the whole positive half-axis. Otherwise this is satisfied for small order statistics. Life distribution (5.85) has a finite support with probability mass $1 - W(\alpha_*)$ at the right endpoint. Below we specify results for distributions with increasing density and failure rate.

Proposition 20 (increasing density) *If $(j-1)/n < 1/\sqrt{3} \approx 0.57735$, then*

$$\frac{E_F Y_{j:n}}{m_F} \leq \frac{n}{n+1-j}\left(1 - \frac{2}{\sqrt{3}}\frac{j-1}{n}\right)^{1/2}. \qquad (5.87)$$

The equality holds for a mixture of the uniform distribution on the interval $[0, m_F/[1 - 2(j-1)/(\sqrt{3}n)]^{1/2}]$ and the Dirac distribution concentrated at $m_F/[1-2(j-1)/(\sqrt{3}n)]^{1/2}]^{1/2}$ with probabilities $\sqrt{3}(j-1)/n$ and $1-\sqrt{3}(j-1)/n$, respectively.

If $(j-1)/n > 1/\sqrt{3}$, then sharp bound

$$\frac{E_F Y_{j:n}}{m_F} \leq \frac{\sqrt{3}}{2}\left(1 + \frac{j-1}{n}\right) \qquad (5.88)$$

is attained by the uniform distribution on the interval $[0, \sqrt{3}m_F]$.

Proposition 21 (increasing failure rate) *For $\alpha_* > \ln[n/(n+1-j)]$ uniquely defined by equation*

$$\frac{2}{\alpha}(1 - e^{-\alpha}) - e^{-\alpha} = \left(1 - \frac{j-1}{n}\right)\left(1 - \ln\frac{n}{n+1-j}\right) \qquad (5.89)$$

and
$$\bar{\nu}^2 = \bar{\nu}_V^2(\alpha_*) = 2[1 - (1-\alpha_*)e^{-\alpha_*}] \tag{5.90}$$

we have
$$\frac{E_F Y_{j:n}}{m_F} \le \frac{n}{n+1-j} \frac{\bar{\nu}_V(\alpha_*)}{\alpha_*}. \tag{5.91}$$

The equality holds iff
$$F(x) = \begin{cases} 0, & \text{if } \frac{x}{m_F} \le 0, \\ 1 - \exp\left(-\bar{\nu}\frac{x}{m_F}\right) & \text{if } 0 \le \frac{x}{m_F} < \frac{\alpha_*}{\bar{\nu}}, \\ 1, & \text{if } \frac{x}{m_F} \ge \frac{\alpha_*}{\bar{\nu}}, \end{cases} \tag{5.92}$$

which is a combination of a right truncated exponential distribution with a pole at the truncation point $m_F \alpha_*/\bar{\nu}$.

5.4 Distributions with Monotone Density and Failure Rate on the Average

The results of this section are based on assertions of Lemmas 8 and 9 for $b = W^{-1}((j-1)/n)$ and $c = d_W$. Therefore one can expect some analogies with inequalities for quantiles of order p derived for $b = W^{-1}(p)$ and $c \searrow b$. Observe that norm (3.57) depends on c, and the normalized projection (3.58) does not. It follows that bounds for quantiles and expected order statistics substantially differ, but they are attained by the same distributions when $p = (j-1)/n$. Note that condition (3.50) is false for $c = d_W$, and there are no trivial bounds $E_F Y_{j:n} \le \mu_F$ for $j \ge 2$. In Theorem 25 and Propositions 21 and 22 we omit explicit descriptions of distributions for which bounds are attained, and refer the reader to respective assertions in Theorem 6 and Propositions 5 and 6 in which p and b should be replaced by $(j-1)/n$ and $W^{-1}((j-1)/n)$, respectively.

Theorem 25 ($F \succeq_* W$) *For* $2 \le j \le n < \infty$, *with* $b = W^{-1}((j-1)/n)$, *and the notation of* (3.20) *and* (3.60), *the following inequality is tight*
$$\frac{E_F Y_{j:n} - \mu_F}{\sigma_F} \le \frac{j-1}{n} \frac{\mu_W(b)}{\vartheta_W(a,b)}. \tag{5.93}$$

Proposition 22 (decreasing density on the average) *For* $2 \le j \le n < \infty$, *bound*
$$\frac{E_F Y_{j:n} - \mu_F}{\sigma_F} \le \frac{\sqrt{3}(j-1)(n+j-1)}{n^2 \theta_U(\frac{j-1}{n})} \tag{5.94}$$

(cf. (3.64)*) is the best possible.*

Proposition 23 (decreasing failure rate on the average) *For $2 \leq j \leq n < \infty$, bound*

$$\frac{E_F Y_{j:n} - \mu_F}{\sigma_F} \leq \frac{(j-1)(1 + \ln \frac{n}{n+1-j})}{n\theta_V(\frac{j-1}{n})} \quad (5.95)$$

(cf. (3.67)) is the best possible.

Theorem 26 ($F \preceq_* W$) *For arbitrary continuous distribution function W with a finite second moment there exists a sequence $F_k \preceq_* W$, $k = 1, 2, \ldots$, of distribution functions that have densities and finite second moments, and satisfy*

$$\lim_{k \to \infty} \sup_{P \in \mathcal{P}_n(F_k)} \frac{E_{F_k} Y_{j:n} - \mu_{F_k}}{\sigma_{F_k}} = \left(\frac{j-1}{n+1-j}\right)^{1/2}. \quad (5.96)$$

In particular, bound (5.96) is sharp for distributions with increasing density and failure rate on the average.

Theorem 26 asserts that general bound (5.13) is attained among $F \preceq_* W$ under mild assumptions on the maximal element W. The proof consists in constructing a sequence of antistarshaped superpositions $F_k^{-1}W - \mu_{F_k}$ (with $F_k^{-1}W(a_W) \searrow -\infty$) that integrate to 0, and tend to $\mathbf{1}_{[W^{-1}((j-1)/n), d_W)}$ in $L^2([a_W, d_W), w(x)dx)$. Actually, we can prove an analogous claim for arbitrary limiting function $\sum_{j=1}^{n} d_j \mathbf{1}_{[W^{-1}((j-1)/n), W^{-1}(j/n))}$ with nondecreasing coefficients d_j, $1 \leq j \leq n$, which means that general bound (5.11) for an arbitrary L-statistic cannot be improved once we restrict the family of distributions to any of the form $\{F : F \preceq_* W\}$.

5.5 Symmetric Unimodal and U-Shaped Distributions

Theorem 27 (symmetric unimodal distributions) *If $0 < (j-1)/n \leq 1/3$, then*

$$\frac{E_F Y_{j:n} - \mu_F}{\sigma_F} \leq \frac{2[n(j-1)]^{1/2}}{3(n+1-j)}. \quad (5.97)$$

This becomes the equality if $Y_1 = \mu$ with probability $1 - 3(j-1)/n$ and is uniformly distributed on $[\mu - \sigma[n/(j-1)]^{1/2}, \mu + \sigma[n/(j-1)]^{1/2}]$ with probability $3(j-1)/n$.

If $1/3 \leq (j-1)/n \leq 2/3$, then

$$\frac{E_F Y_{j:n} - \mu_F}{\sigma_F} \leq \sqrt{3}\frac{j-1}{n}, \quad (5.98)$$

which is the equality iff Y_1 is uniformly distributed on $[\mu - \sqrt{3}\sigma, \mu + \sqrt{3}\sigma]$.

If $(j-1)/n > 2/3$, then

$$\frac{\mathrm{E}_F Y_{j:n} - \mu_F}{\sigma_F} \leq \frac{2}{3}\left(\frac{n}{n+1-j}\right)^{1/2}. \tag{5.99}$$

Here the equality is attained by the mixture of the atom at μ and the uniform distribution on $[\mu - \sigma[n/(n+1-j)]^{1/2}, \mu + \sigma[n/(n+1-j)]^{1/2}]$ with probabilities $3(j-1)/n - 2$ and $3[1 - (j-1)/n]$, respectively.

Theorem 28 (symmetric U-shaped distributions) *If either $(j-1)/n < 1/2 - 1/(2\sqrt{3}) \approx 0.21132$ or $(j-1)/n > 1/2 + 1/(2\sqrt{3}) \approx 0.78868$ then inequality (5.98) holds true with the respective condition for equality.*
For $1/2 - 1/(2\sqrt{3}) < (j-1)/n < 1/2 + 1/(2\sqrt{3})$ bound

$$\frac{\mathrm{E}_F Y_{j:n} - \mu_F}{\sigma_F} \leq \frac{n}{n+1-j}\left(\frac{1}{4} - \frac{1}{\sqrt{3}}\left|\frac{j-1}{n} - \frac{1}{2}\right|\right)^{1/2} \tag{5.100}$$

is sharp, and this becomes the equality for the combination of uniform distribution on

$$\left[\mu - \frac{\sigma}{(1 - 4\sqrt{3}|\frac{j-1}{n} - \frac{1}{2}|)^{1/2}}, \mu + \frac{\sigma}{(1 - 4\sqrt{3}|\frac{j-1}{n} - \frac{1}{2}|)^{1/2}}\right]$$

with probability $2\sqrt{3}|(j-1)/n - 1/2|$ and two atoms at the ends of the interval, with identical probabilities $1/2 - \sqrt{3}|(j-1)/n - 1/2|$.

Proofs of Theorems 27 and 28 are based on projecting folded functionals (5.18) onto convex cones $\mathcal{C}^+_{\succeq_c 2U-1}$ and $\mathcal{C}^+_{\preceq_c 2U-1}$. Observe that

$$g^s_{j:n}(x) = \frac{j-1}{n+1-j} g^s_{n+1-j:n}(x),$$

and the same holds for the respective projections and their norms. This explains the relation between the bounds in both theorems and the fact that the distributions with extreme expectations of jth smallest and largest order statistics are identical. We can immediately conclude similar bounds for some L-statistics.

In contrast with the independent case, all the bounds are nontrivial for any order statistic except that of the sample minimum. Table 5.1 contains numerical values of mean-variance bounds for order statistics from dependent samples of size $n = 20$ coming from the following families of distributions: general (G), symmetric (S), symmetric unimodal (SUN) and U-shaped (SUS), distributions with decreasing density and failure rate (DD and DFR, respectively), and with decreasing density and failure rate on the average (DDA and DFRA, respectively). In fact, all the bounds depend on

5.5 Symmetric Unimodal and U-Shaped Distributions

TABLE 5.1. Sharp uniform mean-variance bounds on expectations of order statistics from dependent samples of size 20 for various families of distributions.

j	G	S	SUN	SUS	DD	DFR	DDA	DFRA	ID	IFR
2	0.22942	0.16644	0.15692	0.08660	0.08660	0.05129	0.09028	0.05250	0.21558	0.21583
3	0.33333	0.24845	0.23424	0.17321	0.17321	0.10536	0.18543	0.10999	0.31208	0.31285
4	0.42008	0.32219	0.30376	0.25981	0.25981	0.16252	0.28232	0.17247	0.39167	0.39325
5	0.5	0.39529	0.37269	0.34641	0.34641	0.22314	0.37897	0.23995	0.46398	0.46673
6	0.57735	0.47141	0.44444	0.43341	0.43301	0.28768	0.47458	0.31254	0.53287	0.53725
7	0.65465	0.55328	0.52164	0.52398	0.51962	0.35668	0.56934	0.39046	0.60045	0.60705
8	0.73380	0.64359	0.60622	0.62188	0.60624	0.43078	0.66414	0.47409	0.66814	0.67779
9	0.81650	0.74536	0.69282	0.73080	0.69389	0.51083	0.76038	0.56408	0.73703	0.75084
10	0.90453	0.86244	0.77942	0.85499	0.78496	0.59784	0.85983	0.66139	0.80802	0.82760
11	1	1	0.86603	1	0.88192	0.69318	0.96490	0.76749	0.88192	0.90962
12	1.10554	1.05409	0.95263	1.04499	0.98758	0.79851	1.07834	0.88455	0.95939	0.99878
13	1.22474	1.11803	1.03923	1.09620	1.10554	0.91629	1.20402	1.01574	1.04083	1.09750
14	1.36277	1.19523	1.12583	1.15493	1.24084	1.04984	1.34731	1.16593	1.12587	1.20916
15	1.52753	1.29099	1.21716	1.22261	1.40106	1.20521	1.51632	1.34278	1.21244	1.33867
16	1.73205	1.41421	1.33333	1.30023	1.59861	1.39393	1.72425	1.55911	1.29904	1.49376
17	2	1.58114	1.49071	1.38564	1.85592	1.63670	1.99448	1.83837	1.38564	1.68751
18	2.38048	1.82574	1.72133	1.47224	2.21944	1.97612	2.37741	2.22936	1.47224	1.94488
19	3	2.23607	2.10819	1.55885	2.80872	2.52143	2.99846	2.85796	1.55885	2.32232
20	4.35890	3.16228	2.98142	1.64545	4.09607	3.70340	4.35840	4.22194	1.64545	2.99772

value $(j-1)/n$ only, because they were derived by means of projecting functions of $(j-1)/n$. Therefore all estimates for $Y_{j:n}$ hold also true for any $Y_{k:m}$ provided that $(k-1)/m = (j-1)/n$. In particular, the moment bounds for the sample medians $Y_{n:2n+1}$ do not change under increase of the sample size. Moreover, one can check that all the bounds for restricted families of distributions presented in this chapter increase continuously in $(j-1)/n$. Consequently, Table 5.1 may provide fair approximations of bounds for $Y_{k:m}$ from various families of parent distributions with $(k-1)/m \approx (j-1)/20$ for some $2 \leq j \leq 20$. Some tables for second moment bounds were presented in Gajek and Rychlik [33]. Also, the third column (SUN) of Table 5.1 comes from that paper. The numerical results for distributions with monotone failure rate and monotone failure rate on the average were earlier presented in Rychlik [87].

5.6 Bias of Quantile Estimates

As in Section 4.5, we consider the problem of evaluating the bias for estimation of population quantiles $F^{-1}(p)$ by sample quantiles $X_{j:n}$, with the only difference being that dependent observations are possible now. The results stated here are valid for $(j-1)/n \leq p < 1$. The problem of establishing the upper bounds on the bias in general populations consists in finding the derivative of the greatest convex minorant of the nonmonotone distribution function

$$G_{j:n}(x) - \mathbf{1}_{[p,1)}(x) = \begin{cases} 0, & \text{if } 0 \leq x \leq \frac{j-1}{n}, \\ \frac{nx+1-j}{n+1-j}, & \text{if } \frac{j-1}{n} \leq x < p, \\ \frac{n(x-1)}{n+1-j}, & \text{if } p \leq x < 1, \end{cases}$$

calculating its L^2-norm, and the form of the normalized derivative. The lower bound amounts to the negative of the upper one for

$$F^{-1}(p) + \sup_{P \in \mathcal{P}_n(F)} E_P(-X_{j:n}) = \int_0^1 F^{-1}(x)(\mathbf{1}_{[p,1)} - G_{-\mathbf{e}_j})(dx),$$

where $\mathbf{e}_j$ is the jth unit vector, and

$$G_{-\mathbf{e}_j}(x) = -\min\left\{\frac{n}{j}x, 1\right\}$$

(cf. (2.25) and (2.26)). Again, we use the Moriguti [58] algorithm based on the greatest convex minorant. The final results are shown below.

Theorem 29 (general distributions) *For all $(j-1)/n \leq p < 1$, we have*

$$-\underline{C}^{\nearrow}(j,n,p) \leq \frac{E_F Y_{j:n} - F^{-1}(p)}{\sigma_F} \leq \bar{C}^{\nearrow}(j,n,p) \tag{5.101}$$

5.6 Bias of Quantile Estimates

with

$$\underline{C}^{\nearrow}(j,n,p) = \left(\frac{1}{1-p} + \frac{n}{j-1}\right)^{1/2},$$

$$\bar{C}^{\nearrow}(j,n,p) = \frac{n}{n+1-j}\left(\frac{1-p}{p}\right)^{1/2}.$$

The upper bound in (5.101) is attained by the two-point marginal distribution (4.106). The lower one is attained by the three-point distribution

$$P\left(X = \mu - \sigma\left[\frac{1-p}{1-p+(j-1)/n}\right]^{1/2}\right) = \frac{j-1}{n},$$

$$P(X = \mu) = p - \frac{j-1}{n},$$

$$P\left(X = \mu + \sigma\left[\frac{(j-1)/n}{1-p+(j-1)/n}\right]^{1/2}\right) = 1-p.$$

Unlike in the independent case, there is no meaningful evaluation of the bias oscillation $\underline{C}^{\nearrow}(j,n,p) + \bar{C}^{\nearrow}(j,n,p)$. However,

$$\frac{\underline{B}^{\nearrow}(j,n,p)}{\underline{C}^{\nearrow}(j,n,p)} = F_{j:n}(p)\left[\frac{j-1}{p(n+j-1-np)}\right]^{1/2} \to \frac{1}{2}, \quad (5.102)$$

$$\frac{\bar{B}^{\nearrow}(j,n,p)}{\bar{C}^{\nearrow}(j,n,p)} = [1 - F_{j:n}(p)]\frac{n+1-j}{n(1-p)} \to \frac{1}{2}, \quad (5.103)$$

as $j/n \to p$, which means that asymptotic maximal bias is twice as great when the independence assumption is violated.

In order to determine the upper bias deviations for populations with $F \succeq_c W$ and $F \succeq_* W$, we need to project the step functions

$$h_{j:n,p,q}(x) = \begin{cases} 0, & \text{if } 0 \leq x < W^{-1}\left(\frac{j-1}{n}\right), \\ \frac{n}{n+1-j} - \frac{1}{q-p}, & \text{if } W^{-1}(p) \leq x < W^{-1}(q), \\ \frac{n}{n+1-j}, & \text{elsewhere,} \end{cases} \quad (5.104)$$

onto the convex cones of nondecreasing convex and nondecreasing star-shaped functions, respectively, in $L^2([a_W, d_W), w(x)dx)$. Fortunately, functions (5.104) obey the assumptions of Lemmas 18 through 21 which provide exact formulae for the projections. As in (4.125),

$$0 < -\int_{\gamma_*}^{c}(x-\gamma_*)h_{j:n,p,q}(x)w(x)\,dx$$

$$= \frac{n(1-q+p)+1-j}{(n+1-j)(q-p)}\int_{\gamma_*}^{c}(x-\gamma_*)w(x)\,dx$$

$$\leq (c-\gamma_*)\frac{n(1-q+p)+1-j}{n+1-j} \to 0,$$

as $q \searrow p$ and $c = W^{-1}(q)$, $\gamma_* \searrow b = W^{-1}(p)$. It follows that the bound for $F \succeq_c W$ has the form of (4.126) with $f_{j:n}W(x)$ replaced by $g_{j:n}W(x) = n\mathbf{1}_{[b,d)}(x)/(n+1-j)$. For $F \succeq_* W$, we have (4.137) with $f_{j:n}W$ replaced by $g_{j:n}W$. In both cases the normalized projection does not depend on the specific form of h. Therefore the upper bounds in the dependent case are attained by the same marginal distributions that provide the analogous bounds for independent observations. Note that the same holds when F is general.

Theorem 30 ($F \succeq_c W$) *If $(j-1)/n \leq p < 1$ and $b = W^{-1}(p)$, then, with notation (4.127) and (4.129), we have*

$$\frac{\mathrm{E}_F Y_{j:n} - F^{-1}(p)}{\sigma_F} \leq \frac{n}{n+1-j} \frac{\hat{\eta}_W(b)}{\hat{\vartheta}_W(b)}. \tag{5.105}$$

Inequality (5.105) is sharp.

Proposition 24 (decreasing density) *If $(j-1)/n \leq p < 1$, then inequality*

$$\frac{\mathrm{E}_F Y_{j:n} - F^{-1}(p)}{\sigma_F} \leq \frac{n}{n+1-j} \left(\frac{3-3p}{1+3p}\right)^{1/2} \tag{5.106}$$

is sharp.

Proposition 25 (decreasing failure rate) *If $(j-1)/n \leq p < 1$, then inequality*

$$\frac{\mathrm{E}_F Y_{j:n} - F^{-1}(p)}{\sigma_F} \leq \frac{n}{n+1-j} \left(\frac{1-p}{1+p}\right)^{1/2} \tag{5.107}$$

is sharp.

Theorem 31 ($F \succeq_* W$) *If $(j-1)/n \leq p < 1$ and $b = W^{-1}(p)$, then, with notation (3.59) and (3.60), we have*

$$\frac{\mathrm{E}_F Y_{j:n} - F^{-1}(p)}{\sigma_F} \leq \bar{C}^0_{\succeq_*W}(j,n,p)$$

$$= \frac{n}{n+1-j} \frac{\eta_W(a,b)}{\vartheta_W(a,b)}. \tag{5.108}$$

Inequality (5.108) is sharp.

Proposition 26 (decreasing density on the average) *If $(j-1)/n \leq p < 1$, then the following inequality is sharp*

$$\frac{\mathrm{E}_F Y_{j:n} - F^{-1}(p)}{\sigma_F} \leq \bar{C}^0_{\succeq_*U}(j,n,p)$$

$$= \frac{n(1+p)}{n+1-j} \left[\frac{3(1-p)}{1+p+7p^2+3p^3}\right]^{1/2}. \tag{5.109}$$

Proposition 27 (decreasing failure rate on the average) *If we have* $(j-1)/n \leq p < 1$, *then the following inequality is sharp*

$$\frac{E_F Y_{j:n} - F^{-1}(p)}{\sigma_F} \leq \bar{C}^0_{\succeq_* V}(j,n,p)$$

$$= \frac{n}{n+1-j} \left\{ \frac{1-p}{1+p[1-\ln(1-p)]^2} \right\}^{1/2}. \quad (5.110)$$

If $j/n \to p$, then bounds (5.105) through (5.107) do not tend to 0, unlike the independent case. However, comparing asymptotic values of the upper bias deviations for $F \succeq_* W, U$, and V we obtain nontrivial results:

$$\frac{\bar{B}^0_{\succeq_* W}(j,n,p)}{\bar{C}^0_{\succeq_* W}(j,n,p)} \to \frac{(b-a) \int_b^d w(x)\, dx}{2 \int_b^d (x-a) w(x)\, dx}$$

$$= \frac{W^{-1}(p) - a_W}{2[\mu_W(W^{-1}(p)) - a_W]} < \frac{1}{2}, \quad (5.111)$$

$$\frac{\bar{B}^0_{\succeq_* U}(j,n,p)}{\bar{C}^0_{\succeq_* U}(j,n,p)} \to \frac{p}{2(1+p)}, \quad (5.112)$$

$$\frac{\bar{B}^0_{\succeq_* V}(j,n,p)}{\bar{C}^0_{\succeq_* V}(j,n,p)} \to \frac{-\ln(1-p)}{2[1-\ln(1-p)]}. \quad (5.113)$$

If $p \searrow 0$, all the limits in (5.111) to (5.113) converge to 0. Differentiating the first one with respect to b, we check that this is increasing. The limiting values of the right-hand sides of (5.112) and (5.113) for $p \nearrow 1$ are $1/4$ and $1/2$, respectively. We recall (5.103) to point out that the respective limits in the general case amount to $1/2$ for all p.

Table 5.2 is the counterpart of Table 4.2 for the dependent case. This enables us to compare respective maximal upper bias deviations for finite samples of size $n = 20$. The effect of dependence is actually notable in all the families. Just as for the independent samples of general populations, the quantile estimation is less precise on the tails. The same concerns the dependent samples with decreasing density functions. For the DFR and DFRA distributions, the upper bias deviations increase in j and p in common with the i.i.d. case.

5.7 Extreme Effect of Dependence

The problem we consider here stems from the robust statistics whose domain is studying sensitivity of statistical procedures against violations of standard assumptions about statistical models. Robustness of L-statistics which are popular tools in robust and nonparametric inference was studied

TABLE 5.2. Sharp uniform variance bounds on upper bias deviations of estimators $Y_{j:20}$ of pth quantiles, $p = j/20$, for various families of distributions (dependent case).

j	G	DD	DFR	DDA	DFRA
1	4.35890	1.57425	0.95119	1.71534	0.99748
2	3.15789	1.51700	0.95214	1.75672	1.04201
3	2.64497	1.47348	0.95925	1.77758	1.08589
4	2.35294	1.44088	0.96058	1.78338	1.12918
5	2.16506	1.41737	0.96825	1.77967	1.17204
6	2.03670	1.40175	0.97840	1.77127	1.21477
7	1.94681	1.39329	0.99127	1.76200	1.25780
8	1.88422	1.39159	1.00716	1.75471	1.30171
9	1.84257	1.39655	1.02647	1.75159	1.34727
10	1.81818	1.40836	1.04973	1.75437	1.39544
11	1.80907	1.42749	1.07763	1.76456	1.44744
12	1.81444	1.45479	1.11111	1.78373	1.50480
13	1.83450	1.49150	1.15142	1.81369	1.56953
14	1.87044	1.53947	1.20024	1.85672	1.64422
15	1.92450	1.60128	1.25988	1.91583	1.73235
16	2	1.68034	1.33333	1.99488	1.83837
17	2.10042	1.78017	1.42374	2.09772	1.96708
18	2.22222	1.89832	1.52944	2.22108	2.11701
19	2.29416	1.97386	1.60128	2.29389	2.22208

5.7 Extreme Effect of Dependence

by many authors (we refer merely to monographs of Huber [41] and Hampel et al. [37] and references given there). For standard i.i.d. parametric models, many violations of marginals were studied. Dependence-robustness for location models was analyzed in Rychlik [80]. The projection method provides tools for determining expectation sensitivity of arbitrary L-statistics against dependence of observations

$$\sup_{P \in \mathcal{P}_n(F)} E_F \sum_{j=1}^n c_j Y_{j:n} - E_F \sum_{j=1}^n c_j X_{j:n}$$

$$= \int_0^1 F^{-1}(x) \left[G_{\mathbf{c}}(x) - \sum_{j=1}^n c_j f_{j:n}(x) \right] dx, \quad (5.114)$$

for various families of parent marginals F.

We present some preliminary results of Rychlik [88], where robustness of single order statistics coming from general populations was studied. To evaluate robustness of the jth order statistic in standard deviation units, we first determine the projection of $h_{j:n} = g_{j:n} - f_{j:n}$ onto $\mathcal{C}^0$ in $L^2([0,1], dx)$. Since $h_{j:n}$ is the difference of density functions, $P^0 h_{j:n} = P^\nearrow h_{j:n}$. Thus we are reduced to establishing projections onto the cone of nondecreasing functions that can be determined by means of the greatest convex minorants. The solution is trivial for the sample minimum, since

$$h_{1:n}(x) = 1 - n(1-x)^{n-1} \quad (5.115)$$

is actually increasing and thus $P^\nearrow h_{1:n} = h_{1:n}$. Another simple solution emerges for the sample maximum. Then

$$h_{n:n}(x) = \begin{cases} -nx^{n-1}, & \text{if } 0 \le x < 1 - \frac{1}{n}, \\ n(1 - x^{n-1}), & \text{if } 1 - \frac{1}{n} \le x \le 1, \end{cases} \quad (5.116)$$

has antiderivative

$$H_{n:n}(x) = \begin{cases} -x^n, & \text{if } 0 \le x \le 1 - \frac{1}{n}, \\ n(x-1) + 1 - x^n, & \text{if } 1 - \frac{1}{n} \le x \le 1, \end{cases} \quad (5.117)$$

which is decreasing on $[0, 1 - 1/n]$, increasing on $[1 - 1/n, 1]$, and concave on both intervals. Therefore

$$\bar{H}_{n:n}(x) = \begin{cases} -\left(1 - \frac{1}{n}\right)^{n-1} x, & \text{if } 0 \le x \le 1 - \frac{1}{n}, \\ n\left(1 - \frac{1}{n}\right)^n (x - 1), & \text{if } 1 - \frac{1}{n} \le x \le 1, \end{cases} \quad (5.118)$$

and

$$P^\nearrow h_{n:n}(x) = \begin{cases} -\left(1 - \frac{1}{n}\right)^{n-1}, & \text{if } 0 \le x < 1 - \frac{1}{n}, \\ n\left(1 - \frac{1}{n}\right)^n, & \text{if } 1 - \frac{1}{n} \le x \le 1. \end{cases} \quad (5.119)$$

A deeper analysis is needed for $2 \leq j \leq n-1$. Then $h_{j:n}$ decreases from 0 at 0 to $-f_{j:n}((j-1)/n)$, jumps up by $n/(n+1-j)$, runs down to $h_{j:n}((j-1)/(n-1))$ where $f_{j:n}$ is maximized and eventually increases to $h_{j:n}(1) = n/(n+1-j)$. It is important to know the sign of $h_{j:n}((j-1)/(n-1))$. It can be shown that this is negative for small and moderate j, and the proportion of js for which this is true increases to 1, as n becomes large. In this case, a thorough analysis shows that either $H_{j:n}((j-1)/n)$ lies above the greatest convex minorant which consists the line tangent to $H_{j:n}$ at 0 and some $s > (j-1)/(n-1)$, and $H_{j:n}$ itself right to s, or $H_{j:n}((j-1)/n)$ spans the minorant which has two linear pieces joining at $(j-1)/n$ and coincides with $H_{j:n}$ in the right part. If $h_{j:n}((j-1)/(n-1)) \geq 0$, then the convex minorant has the form analogous to the latter one, with the only difference being that the slope of the second linear piece is positive. Analytic forms of assumptions and resulting projection are precisely described in Lemma 25.

Lemma 25 *Set*

$$S_{j:n}(x) = \frac{H_{j:n}(x)}{x}, \qquad 0 < x \leq 1, \qquad (5.120)$$

$$T_{j:n}(x) = \frac{H_{j:n}(x) - H_{j:n}(\frac{j-1}{n})}{x - \frac{j-1}{n}}, \qquad \frac{j-1}{n} < x \leq 1. \quad (5.121)$$

If $h_{j:n}((j-1)/(n-1)) < 0$, then there exist unique $r \in ((j-1)/(n-1), 1)$ such that $h_{j:n}(r) = 0$ and $s \in [(j-1)/(n-1), r]$ such that either $s = (j-1)/(n-1)$ if $S_{j:n}((j-1)/(n-1)) \leq h_{j:n}((j-1)/(n-1))$ or s is the solution to $S_{j:n}(x) = h_{j:n}(x)$ otherwise.

If

$$h_{j:n}\left(\frac{j-1}{n-1}\right) < 0 \text{ and } S_{j:n}(s) \leq S_{j:n}\left(\frac{j-1}{n}\right), \qquad (5.122)$$

then

$$P^{\nearrow} h_{j:n}(x) = \begin{cases} S_{j:n}(s), & \text{if } 0 \leq x < s, \\ h_{j:n}(x), & \text{if } s \leq x \leq 1. \end{cases} \qquad (5.123)$$

If

either $h_{j:n}(\frac{j-1}{n-1}) < 0$ and $S_{j:n}(s) > S_{j:n}(\frac{j-1}{n})$ or $h_{j:n}(\frac{j-1}{n-1}) \geq 0$, (5.124)

then there exists a unique $t \in ((j-1)/(n-1), 1)$ such that $T_{j:n}(t) = h_{j:n}(t)$, and

$$P^{\nearrow} h_{j:n}(x) = \begin{cases} S_{j:n}(\frac{j-1}{n}), & \text{if } 0 \leq x < \frac{j-1}{n}, \\ h_{j:n}(t), & \text{if } \frac{j-1}{n} \leq x \leq t, \\ h_{j:n}(x), & \text{if } t \leq x \leq 1. \end{cases} \qquad (5.125)$$

For the prevailing number of cases, we have (5.122) with $s > (j-1)/(n-1)$. Then (5.123) is continuous, and can be written as

$$P^{\nearrow} h_{j:n}(x) = h_{j:n}(\max\{s, x\}).$$

This form is easier to handle in numerical calculations. Function (5.125) has the only jump at $(j-1)/n$. Using projections (5.115), (5.119), (5.123), and (5.125), we are in a position to write optimal bounds for the extreme dependence effect.

Theorem 32 (general distributions) *Bounds*

$$\sup_{P \in \mathcal{P}_n(F)} \frac{E_F Y_{1:n} - E_F X_{1:n}}{\sigma_F} = \Delta = \Delta(1, n) = \frac{n-1}{(2n-1)^{1/2}} \quad (5.126)$$

are attained by the marginal distribution functions

$$F(x) = 1 - \left[\frac{1}{n}\left(1 - \frac{n-1}{(2n-1)^{1/2}} \frac{x-\mu}{\sigma}\right)\right]^{1/(n-1)},$$

$$-(2n-1)^{1/2} \leq \frac{x-\mu}{\sigma} \leq \frac{(2n-1)^{1/2}}{n-1}. \quad (5.127)$$

If (5.122) holds for $2 \leq j \leq n-1$, then

$$\sup_{P \in \mathcal{P}_n(F)} \frac{E_F Y_{j:n} - E_F X_{j:n}}{\sigma_F} = \Delta = \Delta(j, n) \quad (5.128)$$

for

$$\Delta^2 = \frac{1}{s}\left[\frac{ns+1-j}{n+1-j} - F_{j:n}(s)\right]^2$$

$$+ \frac{n^2(1-s)}{(n+1-j)^2} - \frac{2n}{n+1-j}[1 - F_{j:n}(s)]$$

$$+ n\frac{\binom{2j-2}{j-1}\binom{2n-2j}{n-j}}{\binom{2n-1}{n}}[1 - F_{2j-1:2n-1}(s)]. \quad (5.129)$$

Bound (5.128) is attained by

$$F(x) = \begin{cases} 0, & \text{if } \frac{x-\mu}{\sigma} < \frac{S_{j:n}(s)}{\Delta}, \\ s, & \text{if } \frac{S_{j:n}(s)}{\Delta} \leq \frac{x-\mu}{\sigma} \leq \frac{h_{j:n}(s)}{\Delta}, \\ h_{j:n}^{-1}\left(\Delta \frac{x-\mu}{\sigma}\right), & \text{if } \frac{h_{j:n}(s)}{\Delta} \leq \frac{x-\mu}{\sigma} \leq \frac{n/(n+1-j)}{\Delta}, \\ 1, & \text{if } \frac{x-\mu}{\sigma} \geq \frac{n/(n+1-j)}{\Delta}. \end{cases} \quad (5.130)$$

5. Order Statistics of Dependent Observations

Under conditions (5.124) we have (5.130) with

$$\Delta^2 = \frac{n}{j-1} F_{j:n}^2\left(\frac{j-1}{n}\right) + \left(t - \frac{j-1}{n}\right)\left[\frac{n}{n+1-j} - f_{j:n}(t)\right]^2$$
$$+ \frac{n^2(1-t)}{(n+1-j)^2} - \frac{2n}{n+1-j}[1 - F_{j:n}(t)]$$
$$+ n\frac{\binom{2j-2}{j-1}\binom{2n-2j}{n-j}}{\binom{2n-1}{n}}[1 - F_{2j-1:2n-1}(t)]. \tag{5.131}$$

The supremum is attained by (5.130) with s, $S_{j:n}(s)$, and $h_{j:n}(s)$ replaced by $(j-1)/n$, $S_{j:n}((j-1)/n)$ and $h_{j:n}(t)$, respectively.

Finally, the bound

$$\sup_{P \in \mathcal{P}_n(F)} \frac{E_F Y_{n:n} - E_F X_{n:n}}{\sigma_F} = \frac{(n-1)^{n-1/2}}{n^{n-1}}, \tag{5.132}$$

is attained by the two-point marginal distribution concentrated on points $\mu - \sigma/(n-1)^{1/2}$ and $\mu + \sigma(n-1)^{1/2}$ with respective probabilities $1 - 1/n$ and $1/n$.

In the problem of measuring the effect of dependence, it is of vital interest to evaluate the lower deviations

$$\Delta^-(j,n) = \sup_{P \in \mathcal{P}_n(F)} \frac{EX_{j:n} - EY'_{j:n}}{\sigma_F}.$$

Fortunately, once we consider the class of arbitrary marginals, we can get an immediate answer. Indeed, introducing

$$X_j^- = 2\mu_F - X_j,$$

$1 \leq j \leq n$, with common distribution function

$$F^-(x) = 1 - F(2\mu_F - x-)$$

and order statistics

$$X_{j:n}^- = 2\mu_F - X_{n+1-j:n}$$

(with the same notation for Ys), we have

$$\Delta^-(j,n) = -\inf_{F \in \mathcal{F}_p} \frac{EY_{j:n} - EX_{j:n}}{\sigma_F}$$
$$= \sup_{F^- \in \mathcal{F}} \frac{E_{F^-}Y_{n+1-j:n}^- - E_{F^-}X_{n+1-j:n}^-}{\sigma_F}$$
$$= \Delta(n+1-j, n). \tag{5.133}$$

Accordingly, we conclude that the lower bound on the deviation of the expectation of the jth smallest order statistic under dependence is identical with the upper bound on the deviation of the jth largest one. Similar arguments explain the coincidence of (5.126) with the mean-variance bound for the maximum in the i.i.d. sample (see (4.2)), and the fact that these bounds are attained by distribution functions of variables mutually symmetric about μ (see (4.3) and (5.127)). This is a consequence of relations

$$\sup_{P \in \mathcal{P}_n(F)} E_F Y_{1:n} - E_F X_{1:n} = \mu_F - E_F X_{1:n} = E_{F^-} X_{1:n}^- - \mu_{F^-}. \quad (5.134)$$

In fact, introducing dependence of X_j^- on μ_F, being necessary in (5.134), is redundant in arguing that leads to (5.133). Comparing (5.13) and (5.14) for $j = n$ with (5.132) we observe that

$$\frac{\sup E_F Y_{n:n} - E_F X_{n:n}}{\sup E_F Y_{n:n} - \mu_F} = \left(1 - \frac{1}{n}\right)^{n-1} \searrow e^{-1} \approx 0.36788, \quad (5.135)$$

and both the suprema in (5.135), taken over all $P \in \mathcal{P}_n(F)$ are achieved by the same distribution function.

Numerical calculations based on Theorem 32 show that for general marginal distributions central order statistics are more robust under dependence than the extreme ones. We can also see that each jth smallest order statistic is more sensitive than the respective jth greatest one. In fact, Rychlik [88] measured dependence-robustness of order statistics in terms of scale parameters generated by central absolute moments of order $1 \leq p \leq \infty$. Numerical analysis shows similarity of conclusion for cases $p = 1$ and $p = 2$, which were presented above. For $p = +\infty$ (i.e., for bounded observations), the conclusions are diametrically opposite. An interesting fact to note here is that (5.135) holds true for the classes of marginal distributions with finite pth moment for arbitrary $1 \leq p \leq \infty$.

5.8 Open Problems

1. What are the second moment bounds on order statistics of dependent samples from populations determined by the star order? It is of particular interest for $F \preceq_* W$ for which the mean-variance bounds coincide with the general ones. In the case of life distributions, conditions of attainability require unbounded increase of the second raw moments, which seems to be unrealistic in many practical problems.

2. From the results for single order statistics we can immediately conclude bounds for some L-statistics. For instance, we have

$$\sup_{P \in \mathcal{P}_n(F)} \mathrm{E}_F Y_{j:n} = \int_0^1 F^{-1}(x) g_{j:n}(x)\, dx$$

$$= \sup_{P \in \mathcal{P}_n(F)} \mathrm{E}_F \frac{1}{k+1-j} \sum_{l=j}^{k} Y_{l:n} \qquad (5.136)$$

$$= \sup_{P \in \mathcal{P}_n(F)} \mathrm{E}_F \left[\frac{j}{n} Y_{j:n} + \frac{1}{n} \sum_{l=j+1}^{k-1} Y_{l:n} \right.$$
$$\left. + \frac{n+1-k}{n} Y_{k:n} \right], \qquad (5.137)$$

$1 \leq j \leq k \leq n$, (cf. (2.25) and (2.26)), and therefore all the bounds derived in this chapter for a given jth order statistic coincide with the bounds for trimmed and Winsorized means ((5.136) and (5.137), respectively) that reject $j-1$ smallest observations. Moreover, for $2 \leq j \leq n$, we have

$$\sup_{P \in \mathcal{P}_n(F)} \mathrm{E}_F (Y_{j:n} - Y_{j-1:n}) = \frac{n}{j-1} \int_0^1 F^{-1}(x)[g_{j:n}(x) - 1]\, dx$$

$$= \frac{n}{j-1} \sup_{P \in \mathcal{P}_n(F)} (\mathrm{E}_F Y_{j:n} - \mu_F).$$

Therefore we obtain bounds for spacings $Y_{j:n} - Y_{j-1:n}$ in standard deviation units by multiplying all respective mean-variance bounds for single jth order statistics of various families of sample distributions by factor $n/(j-1)$. In all these cases, the bounds are attained by the same joint distributions. This is not so for the differences of other pairs of order statistics. Then, due to (2.25) and (2.26), we have

$$\sup_{P \in \mathcal{P}_n(F)} \mathrm{E}_F(Y_{j:n} - Y_{i:n})$$
$$= \int_0^1 F^{-1}(x) \left[\frac{n}{n+1-j} \mathbf{1}_{((j-1)/n,1)}(x) - \frac{n}{i} \mathbf{1}_{[0,i/n)}(x) \right] dx$$

when $2 \leq i+1 < j \leq n$. Therefore we should cope with the projections of three-valued step functions onto various convex cones. Gajek and Rychlik [32] proved that the projection of an n-valued step function (which corresponds with general L-statistics of dependent samples of size n at least) onto $\mathcal{C}_{\succeq_c W}^+$ is a broken line with n pieces at most. We suspect a similar conclusion for projections onto other cones determined by the convex order. However, it is a difficult analytic problem to find the best approximations in such large parametric classes.

5.8 Open Problems

3. Studies of bias in quantile estimation in the independent and dependent samples were carried out for the same families of marginals. Formulating unsolved problems in the latter case, we can directly refer to the last two open problems of the previous chapter, where other classes of interest were indicated. It seems that projection problems may be solved more easily here, because the functional has a simpler form.

4. We have no evaluations of the dependence effect on order statistics from restricted families of marginal distributions. For some families (e.g., those of symmetric and symmetric unimodal distributions) the lower bounds could be deduced from the upper ones. We also deduce the lower bounds for distributions with increasing density once we find the upper bounds for distributions with decreasing density, and vice versa. However, this is not so if we take into account distributions with monotone failure rate, and many other pairs of mutually dual families. We also mention the importance of analogous evaluations of robustness against dependence for many popular L-statistics: spacings, ranges, trimmed means, and the like.

6
Records and kth Records

In comparison with evaluations of other statistical functionals discussed here, investigations for record values are still at a preliminary stage, and only a few results are presented now. Examples of Section 6.1 show that the range of record values can be arbitrarily large when all types of interdependence among the original variables are admitted. For the case of i.i.d. sequences with general and symmetric distributions, mean-variance bounds on standard and kth records, due to Nagaraja [59] and Raqab [74], respectively, are presented in Section 6.2. In Section 6.3 second moment bounds for distributions with decreasing density and failure rate are cited from Gajek and Okolewski [31]. Finally, we discuss evaluations of Rychlik [83] for increments of first records coming from various families of parent distributions. Note that Raqab [75] derived pth absolute moment bounds on expectations of first records in general and symmetric populations based on the Hölder inequality. Nagaraja [59] used the Jensen inequality for deriving some quantile bounds on expectations of records (see also Arnold and Balakrishnan [5, Section 6.2]). Some bounds and approximations can be found in Arnold et al. [8, Sections 3.8 and 3.9].

6.1 Dependent Identically Distributed Observations

The results of the previous chapter prove that numerous nontrivial inequalities can be established for expectations of order statistics under the

132   6. Records and kth Records

assumption that all observations have an identical distribution, but they are arbitrarily dependent. However, the assumption admits probabilistic models with peculiar properties of records. The examples presented below show that stating the problem of bounds for the expectation of record values of arbitrarily dependent observations makes no sense.

In the sequence of identical variables $Y_1 = Y_2 = \ldots$, the primary record value $R_0 = Y_1$ can never be improved, and kth records cannot be defined for $k \geq 2$. On the other hand, we can construct a sequence of identically distributed variables for which the first record value may be arbitrarily large. Precisely, the notion of being arbitrarily large depends here on properties of distribution F of single observations. If F has an atom at the right endpoint of its support d_F, then $R_1 = d_F$. If d_F is finite and F is continuous at d_F, then R_1 can be arbitrarily close to d_F. If d_F is infinite, then R_1 may have an arbitrarily large value. The conclusions follow from the next theorem.

Theorem 33 *For an arbitrary positive integer n, there exists a sequence of standard uniform random variables Y_j, $j \geq 1$, for which $R_1 \geq 1 - 1/n$.*

The proof is constructive. Combining the construction with transformation $F^{-1}(Y_j)$, $j \geq 1$, we obtain a sequence of dependent random variables with common distribution function F whose first record value satisfies $R_1 \geq F^{-1}(1 - 1/n)$. This is arbitrarily large in the sense described above.

PROOF. Fix n, and take n random variables U_i, which are uniformly distributed on intervals $[(i-1)/n, i/n]$, $1 \leq i \leq n$. Here U_i may be independent or arbitrarily dependent, for example generated for a single uniform random variable U_0 by means of transformations

$$U_i = \frac{i - 1 + U_0}{n}, \qquad 1 \leq i \leq n.$$

Now create n random vectors

$$\mathbf{U}_i = (U_i, U_{i-1}, \ldots, U_1, U_n, U_{n-1}, \ldots, U_{i+1}), \qquad 1 \leq i \leq n. \qquad (6.1)$$

Let $(Y_1, \ldots, Y_n)$ coincide with a randomly chosen vector $\mathbf{U}_i$. We easily check that Y_j, $1 \leq j \leq n$, are uniformly distributed on $[0, 1]$. Now extend the sequence by adding i.i.d. random variables Y_j, $j > n$, with the same uniform distribution, and independent of $(Y_1, \ldots, Y_n)$. If $(Y_1, \ldots, Y_n) = \mathbf{U}_i$ for $1 \leq i < n$, then $L_1 = i + 1$, and

$$R_1 = U_n \in \left[1 - \frac{1}{n}, 1\right].$$

In order to complete the proof we notice that otherwise $L_1 > n$ and

$$R_1 \in (U_n, 1] \subset \left[1 - \frac{1}{n}, 1\right]. \qquad \blacksquare$$

Analogous assertions can be proved for general kth records. It suffices to replace single U_i by k-tuples of (independent, say) variables uniformly distributed on $[(i-1)/n, i/n]$, $1 \leq i \leq n$, generating vectors $\mathbf{U}_i$ (see (6.1)) of length kn. There were some attempts at imposing stronger conditions on the dependence structure of observations (e.g., exchangeability in Arnold and Balakrishnan [5, Section 6.3], stationary Markov property in Biondini and Siddiqui [16]), but they do not lead to representations for which our projection method works. In the remainder of this chapter we therefore confine ourselves to record values of independent sequences.

6.2 General and Symmetric Distributions

Bounds on expectations of kth records for general and symmetric distributions were analyzed in Raqab [74] by means of Moriguti [58] projection of functions (2.40) based on greatest convex minorants. Special cases of first records for which (2.40) are convex functions varying from 0 to ∞, were solved in Nagaraja [59] by use of the Schwarz inequality. For general F, we therefore have

$$\frac{E_F R_n - \mu_F}{\sigma_F} \leq \left\{ \frac{1}{(n!)^2} \int_0^1 [-\ln(1-x)]^{2n}\, dx - 1 \right\}^{1/2}$$

$$= \left[\binom{2n}{n} \int_0^1 f_{2n}(x)\, dx - 1 \right]^{1/2}$$

$$= \left[\binom{2n}{n} - 1 \right]^{1/2}, \quad (6.2)$$

$n \geq 1$, which becomes the equality for an affine transformation of the Weibull distribution

$$F(x) = 1 - \exp\left(-\left[n!\left(1 + D\frac{x-\mu}{\sigma}\right)\right]^{1/n}\right) \mathbf{1}_{[\mu-\sigma/D,\infty)}(x) \quad (6.3)$$

with shape parameter $1/n$ and

$$D = D^0(1,n) = \left[\binom{2n}{n} - 1 \right]^{1/2}.$$

Observe that (6.3) has decreasing density and failure rate.

If $k \geq 2$, then (2.40) vanishes at 0 and 1, and increases on $[0, V(n/(k-1))]$, and ultimately decreases. Analysis similar to that in Section 4.1 yields

$$P^0(f_n^{(k)} - \mathbf{1})(x) = P^{\nearrow} f_n^{(k)}(x) - 1 = \bar{f}_n^{(k)}(x) - 1$$
$$= f_n^{(k)}(\min\{x, \alpha_*\}) - 1,$$

where α_* is uniquely defined in the following theorem.

Theorem 34 (general distributions) *For given $k \geq 2$ and $n \geq 1$, define $\alpha_* = \alpha_*(k,n) \in (0, V(n/(k-1)))$ as the solution to*

$$(1-x)f_n^{(k)}(x) = 1 - F_n^{(k)}(x)$$
$$= (1-x)^k \sum_{j=0}^{n} \frac{k^j}{j!}[-\ln(1-x)]^j, \quad (6.4)$$

where $f_n^{(k)}(x)$ and $F_n^{(k)}(x)$ are the density function defined in (2.40) and the respective distribution function. Then

$$\frac{E_F R_n^{(k)} - \mu_F}{\sigma_F} \leq D = D^0(k,n) \quad (6.5)$$

for

$$D^2 = \|\bar{f}_n^{(k)}\|^2 - 1$$
$$= k\binom{2n}{n}\left(\frac{k}{2k-1}\right)^{2n+1} F_{2n}^{(2k-1)}(\alpha_*)$$
$$+ (1-\alpha_*)[f_n^{(k)}(\alpha_*)]^2 - 1. \quad (6.6)$$

The equality in (6.5) holds for

$$F(x) = \begin{cases} 0, & \text{if } \frac{x-\mu}{\sigma} \leq -\frac{1}{D}, \\ (f_n^{(k)})^{-1}\left(1 + D\frac{x-\mu}{\sigma}\right), & \text{if } -\frac{1}{D} \leq \frac{x-\mu}{\sigma} < \frac{f_n^{(k)}(\alpha_*)-1}{D}, \\ 1, & \text{if } \frac{x-\mu}{\sigma} \geq \frac{f_n^{(k)}(\alpha_*)-1}{D}. \end{cases} \quad (6.7)$$

For $n = 2, 3$, Equation (6.4) is solved by

$$\alpha_*(k,2) = 1 - \exp\left(-\frac{1}{k(k-1)}\right),$$
$$\alpha_*(k,3) = 1 - \exp\left(-\frac{1 + (2k-1)^{1/2}}{k(k-1)}\right),$$

respectively, and then (6.6) has complicated explicit representations. Distribution function (6.7) has a finite support with a smooth density function, and a pole with probability $1 - \alpha_*$ at the right end.

Analogous second moment bounds have slightly simpler forms

$$\frac{E_F R_n}{m_F} \leq \binom{2n}{n}^{1/2}, \quad n \geq 1, \quad (6.8)$$

$$\frac{E_F R_n^{(k)}}{m_F} \leq \|\bar{f}_n^{(k)}\|, \quad n \geq 1, \, k \geq 2 \quad (6.9)$$

6.2 General and Symmetric Distributions

(cf. (6.6)), and are attained by

$$F(x) = 1 - \exp\left(-\left[\sqrt{(2n)!}\,\frac{x}{m}\right]^{1/n}\right)\mathbf{1}_{[0,\infty)}(x), \tag{6.10}$$

$$F(x) = \begin{cases} 0, & \text{if } \frac{x}{m} \leq 0, \\ (f_n^{(k)})^{-1}(\|\bar{f}_n^{(k)}\|\frac{x}{m}), & \text{if } 0 \leq \frac{x}{m} < \frac{f_n^{(k)}(\alpha_*)}{\|\bar{f}_n^{(k)}\|}, \\ 1, & \text{if } \frac{x}{m} \geq \frac{f_n^{(k)}(\alpha_*)}{\|\bar{f}_n^{(k)}\|}, \end{cases} \tag{6.11}$$

respectively. Both (6.10) and (6.11) define life distributions. Formally, inequalities (6.8) and (6.9) hold true for $n = 0$, and they amount to 1 then. They are attained by degenerate random variables concentrated at $m_F > 0$. Obviously, this can be approximated by life random variables in the mean square with an arbitrarily desired accuracy.

Since
$$s_n(x) = f_n(x) - f_n(1-x), \quad x \in \left[\tfrac{1}{2}, 1\right),$$
is the difference of increasing and decreasing functions with identical values at $1/2$, this is strictly increasing from 0 to ∞. Evaluating expectations of the first records in symmetrically distributed sequences we simply use the Schwarz inequality

$$E_F R_n - \mu_F \leq \int_{1/2}^{1} [F^{-1}(x) - \mu]s_n(x)\,dx$$
$$\leq D^s(1,n)\sigma_F = D\sigma_F, \tag{6.12}$$

where

$$2D^2 = \int_{1/2}^{1} s_n^2(x)\,dx$$
$$= \binom{2n}{n} - \frac{1}{(n!)^2}\int_0^1 \ln^n x \ln^n(1-x)\,dx. \tag{6.13}$$

The bound in (6.12) is attained by

$$F(x) = s_n^{-1}\left(\sqrt{2}D\,\frac{x-\mu}{\sigma}\right), \tag{6.14}$$

which has a smooth symmetric density with the infinite support and symmetry center μ.

Raqab [74] also considered kth records in symmetrically distributed populations using the greatest convex minorant approach. He proposed numerical evaluations that substantially improved bounds for general populations (6.5) and (6.6), and nonsharp ones derived in Grudzień and Szynal [35] based on direct application of the Schwarz inequality. However, Raqab's paper lacks a theoretical justification for use of specific constructions of the greatest convex minorant and precise description of the scope of applicability. Therefore we do not present details here.

6.3 Life Distributions with Decreasing Density and Failure Rate

We now aim at describing the bounds on $E_F R_n^{(k)}/m_F$ for the i.i.d. sequences with the life distributions with decreasing density and failure rate. Therefore we need to project (2.40) and their compositions with the exponential distribution function

$$f_n^{(k)} V(x) = \frac{k^{n+1}}{n!} x^n e^{-(k-1)x}, \quad k \geq 1, \, n \geq 0, \qquad (6.15)$$

onto convex cones $\mathcal{C}_{\succeq_c U}^+$ and $\mathcal{C}_{\succeq_c V}^+$, respectively. We first exclude trivial cases. If $n = 0$, then

$$E_F R_0^{(k)} = E_F X_{1:k} \leq \mu_F \leq m_F$$

for arbitrary F. Referring to the projection method, we obtain the same conclusion. For $n = 0$ both (2.40) and (6.15) are nonincreasing concave functions. It easily follows that their projections are constants amounting to 1 by Lemma 1, with norm 1. We refer to (4.19) for construction of sequences that attain the bound in the limit. If $k = 1$ and $n \geq 1$, then both (2.40) and (6.15) are convex increasing, and so belong to $\mathcal{C}_{\succeq_c U}^+$ and $\mathcal{C}_{\succeq_c V}^+$, respectively. This implies that evaluations of respective expectation functionals by the Schwarz inequality provide the sharp bounds in both cases coinciding with (6.8) valid for arbitrary F. Another way of verifying the sharpness of (6.8) in the classes under study consists in checking that distribution function (6.10) providing equality in (6.8) has decreasing density and failure rate.

In the remaining cases $k \geq 2$ with $n \geq 1$ we make use of auxiliary results of Section 4.2. First we check that the pairs

$$\begin{aligned}(h(x), w(x)) &= (f_n^{(k)}(x), \mathbf{1}_{[0,1)}(x)), \\ (h(x), w(x)) &= (f_n^{(k)} V(x), e^{-x} \mathbf{1}_{[0,\infty)}(x))\end{aligned}$$

obey conditions (4.29) to (4.33). In fact, only verification of (4.33) is nontrivial here. An easy computation shows that (6.15) is convex increasing on $(a, b) = (0, (n-\sqrt{n})/(k-1))$, concave increasing on $(b, c) = ((n-\sqrt{n})/(k-1), n/(k-1))$, and decreasing on $(c, d) = (n/(k-1), \infty)$. Note that $a = b$ for the first record values. More effort is needed for establishing similar conclusions for (2.40). Namely, this is concave increasing on $(0, V(1/(k-1)))$ and decreasing on $(V(1/(k-1)), 1)$ for $n = 1$ and all $k \geq 2$. If $n \geq 2$ and $k \geq 2$, then $f_n^{(k)}$ is convex increasing on $(0, b)$, concave increasing on (b, c), and decreasing on $(c, 1)$ for $c = V(n/(k-1))$ and

$$b = \begin{cases} V(n-1), & \text{if } k = 2, \\ V\left(\frac{n(2k-3)+[n^2+4n(k-1)(k-2)]^{1/2}}{2(k-1)(k-2)}\right), & \text{if } k \geq 3, \end{cases} \qquad (6.16)$$

6.3 Life Distributions with Decreasing Density and Failure Rate

This enables us to apply Lemma 13 to establish the parametric forms of projections of $f_n^{(k)}$ and $f_n^{(k)}V$. Since we have no convex increasing pieces for $n = 1$, the projections are then linear. Consecutive reasoning steps are similar to those in the proof of bounds on order statistics. The main difference is that calculating (4.41) and (4.42) now we do not deal with combinations of Bernstein polynomials, but obtain combinations of density functions $f_m^{(j)}$ with the same j and various m. Without going into details, we try to convince the reader by presenting formulae for derivatives and indefinite integrals of $f_n^{(k)}$ and $f_n^{(k)}V$:

$$(f_n^{(k)})'(x) = \frac{f_{n-1}^{(k)}(x) - (k-1)f_n^{(k)}(x)}{1-x},$$

$$\int_x^1 f_n^{(k)}(y)\,dy = \frac{1-x}{k} \sum_{m=0}^n f_m^{(k)}(x),$$

$$(f_n^{(k)}V)'(x) = f_{n-1}^{(k)}V(x) - (k-1)f_n^{(k)}V(x),$$

$$\int_x^\infty f_n^{(k)}V(y)\,dy = \frac{e^{-x}}{k} \sum_{m=0}^n f_m^{(k)}V(x).$$

It is possible but unnecessary to replace positive functional coefficients in the right-hand sides by more complicated constants with simultaneous replacement of k by neighboring integers. Gajek and Okolewski [31] established the following variation diminishing property for the combinations of density functions of kth record values in the i.i.d. uniform sequences.

Lemma 26 *For every positive integer j and m, the number of zeros in $(0,1)$ of a given nonzero linear combination*

$$f_\mathbf{a}(x) = \sum_{m=0}^n a_m f_m^{(j)}(x), \qquad x \in (0,1),$$

does not exceed the number of sign changes of the sequence $\mathbf{a} = (a_0, \ldots, a_n)$. The first and the last signs of the sum are identical to the signs of the first and last nonzero elements of $\mathbf{a}$, respectively.

Note that it suffices to prove the claim for combinations of

$$g(x; j, m+1) = f_m^{(j)}V(x)e^{-x} = \frac{j^{m+1}}{m!}x^m e^{-jx}\mathbf{1}_{[0,\infty)}(x) \qquad (6.17)$$

which are the density functions of $\Gamma(j, m+1)$ gamma distributions. This follows from the fact that for fixed j, the one-parametric family of gamma densities $g(x; j, \nu)$, $\nu > 0$, is strictly totally positive of order ∞; that is, for every positive integer i, and arbitrary nondecreasing sequences $0 < x_1 < \ldots < x_i$, $0 < \nu_1 < \ldots < \nu_i$, we have

$$\det[g(x_k; j, \nu_l)]_{k,l=1,\ldots,i} > 0$$

(see, e.g., Karlin [44, Chapter 1]). The variation diminishing property for linear combinations of density functions of totally positive families is stated in Karlin and Studden [45, Corollary 4.1].

A thorough analysis of (4.41) and (4.42) enables us to establish bounds described in the following two theorems. We point out apparent analogies with Theorems 11 and 12. Here we take into account $k \geq 2$ and $n \geq 1$, as the remaining cases $k = 1$ and $n = 0$ were discussed above.

Theorem 35 (decreasing density) *If $(1 + 1/k)^{n+1} \leq 3$, then*

$$\frac{E_F R_n^{(k)}}{m_F} \leq \sqrt{3}\left[1 - \left(\frac{k}{k+1}\right)^{n+1}\right], \quad (6.18)$$

which is the equality if F is the uniform distribution on $[0, \sqrt{3}m_F]$. Otherwise, with notations (4.34) and (4.41), we have

$$\frac{E_F R_n^{(k)}}{m_F} \leq D = D_{\succeq_c U}^+(k,n) = \|(f_n^{(k)})_{\alpha_* \beta_*}\|, \quad (6.19)$$

for

$$\begin{aligned}
\|(f_n^{(k)})_{\alpha_* \beta_*}\|^2 &= k\left(\frac{k}{2k-1}\right)^{2n+1}\binom{2n}{n}F_{2n}^{(2k-1)}(\beta_*) \\
&+ (1-\beta_*)[f_n^{(k)}(\beta_*)]^2 + (1-\beta_*)^2 \alpha_* f_n^{(k)}(\beta_*) \\
&+ (1-\beta_*)^3 \alpha_*^2/3,
\end{aligned} \quad (6.20)$$

$$\begin{aligned}
\alpha_* = \alpha_*(\beta_*) &= \frac{3}{(1-\beta_*)^3}\Big\{(1-\beta_*)[1-F_n^{(k)}(\beta_*)] \\
&- \left(\frac{k}{k+1}\right)^{n+1}[1-F_n^{(k+1)}(\beta_*)]\Big\} \\
&- \frac{3}{2(1-\beta_*)}f_n^{(k)}(\beta_*),
\end{aligned} \quad (6.21)$$

where $\beta_* < b$ (the first inflection point of $f_n^{(k)}$ defined in (6.16)) is the smaller of the smallest positive zeros of combinations

$$\begin{aligned}
K_U(x) &= \frac{3}{k}\sum_{m=0}^{n-1}\left[1 - \left(\frac{k}{k+1}\right)^{n-m}\right]f_m^{(k)}(x) \\
&- k f_{n-2}^{(k)}(x) + \left(k - \frac{5}{2}\right)f_{n-1}^{(k)}(x),
\end{aligned} \quad (6.22)$$

6.3 Life Distributions with Decreasing Density and Failure Rate

$$L_U(x) = \frac{1}{2k} \sum_{m=0}^{n-2} \left[3\left(\frac{k}{k+1}\right)^{n-m} - 1 \right] f_m^{(k)}(x)$$
$$- \frac{(k-1)(k-2)}{4k(k+1)} f_{n-1}^{(k)}(x). \tag{6.23}$$

Inequality (6.19) becomes the equality if

$$F(x) = \begin{cases} 0, & \text{if } \frac{x}{m} \leq 0, \\ (f_n^{(k)})^{-1}\left(D\frac{x}{m}\right), & \text{if } 0 \leq \frac{x}{m} \leq \frac{f_n^{(k)}(\beta_*)}{D}, \\ \beta_* + \frac{D\frac{x}{m} - f_n^{(k)}(\beta_*)}{\alpha_*}, & \text{if } \frac{f_n^{(k)}(\beta_*)}{D} \leq \frac{x}{m} \leq \frac{f_n^{(k)}(\beta_*) + \alpha_*(1-\beta_*)}{D}, \\ 1, & \text{if } \frac{x}{m} \geq \frac{f_n^{(k)}(\beta_*) + \alpha_*(1-\beta_*)}{D}. \end{cases} \tag{6.24}$$

Theorem 36 (decreasing failure rate) *If $n \leq 2k - 1$, then*

$$\frac{E_F R_n^{(k)}}{m_F} \leq \frac{n+1}{\sqrt{2k}}, \tag{6.25}$$

which becomes the equality for the exponential distribution with the scale parameter $m_F/\sqrt{2}$.

Otherwise, under notations (4.34) and (4.41), we have

$$\frac{E_F R_n^{(k)}}{m_F} \leq D = D_{\succeq_c V}^+(k, n) = \|(f_n^{(k)} V)_{\alpha_* V^{-1}(\gamma_*)}\|, \tag{6.26}$$

for

$$\|(f_n^{(k)} V)_{\alpha_* V^{-1}(\gamma_*)}\|^2 = (1 - \gamma_*)\{2\alpha_*^2 + 2\alpha_* f_n^{(k)}(\gamma_*) + [f_n^{(k)}(\gamma_*)]^2\}$$
$$+ k\left(\frac{k}{2k-1}\right)^{2n+1} \binom{2n}{n} F_{2n}^{(2k-1)}(\gamma_*), \tag{6.27}$$

$$\alpha_* = \alpha_*(V^{-1}(\gamma_*)) = \frac{n+1}{2k(1-\gamma_*)}[1 - F_{n+1}^{(k)}(\gamma_*)]$$
$$- \frac{V^{-1}(\gamma_*)}{2(1-\gamma_*)}[1 - F_n^{(k)}(\gamma_*)] - \frac{1}{2}f_n^{(k)}(\gamma_*), \tag{6.28}$$

where $\gamma_* < V((n - \sqrt{n})/(k-1))$ is the minimum of the smallest positive zeros of

$$K_V(x) = \sum_{m=0}^{n-3} \frac{n-m}{2k^2} f_m^{(k)}(x) - \frac{1-k^3}{k^2} f_{n-2}^{(k)}(x)$$
$$+ \frac{(k-1)^2(2k-1)}{2k^2} f_{n-1}^{(k)}(x), \tag{6.29}$$

$$L_V(x) = \sum_{m=0}^{n-2} \frac{2k-n+m}{2k^2} f_m^{(k)}(x) - \frac{(k-1)^2}{2k^2} f_{n-1}^{(k)}(x). \quad (6.30)$$

The equality holds in (6.26) if

$$F(x) = \begin{cases} 0, & \text{if } \frac{x}{m} \leq 0, \\ (f_n^{(k)})^{-1}\left(D\frac{x}{m}\right), & \text{if } 0 \leq \frac{x}{m} \leq \frac{f_n^{(k)}(\gamma_*)}{D}, \\ 1 - (1-\gamma_*) \exp\left(-\frac{D\frac{x}{m} - f_n^{(k)}(\gamma_*)}{\alpha_*}\right), & \text{if } \frac{x}{m} \geq \frac{f_n^{(k)}(\gamma_*)}{B}. \end{cases} \quad (6.31)$$

We see that the extreme expectations of kth record values are attained by the uniform and exponential distributions in Theorems 35 and 36, respectively, if n is small. This is true for $n = 1$ and all k in particular, which is an immediate consequence of the linearity of respective projections implied by lacking a convex piece in the left parts of $f_1^{(k)}(x)$ and $f_1^{(k)}V(x)$. Distribution functions $F_m^{(j)}(x)$ of the jth record values appearing in (6.20), (6.21), (6.27), and (6.28) have alternative representations

$$F_m^{(j)}(x) = G(-\ln(1-x); j, m+1),$$

where the right-hand side denotes the gamma distribution functions with densities defined in (6.17), composed with the standard exponential quantile function. Absolutely continuous distribution functions (6.24) and (6.31) have the same form up to a multiplicative factor on the left, and uniform and exponential right tails, respectively. The contributions of the latter amount to $1 - \beta_*$ and $1 - \gamma_*$, respectively.

6.4 Increments of Records

Here we present some standard deviation bounds on expectations of record increments

$$E_F(R_n - R_{n-1}) = \int_0^1 [F^{-1}(x) - \mu][f_n(x) - f_{n-1}(x)]\, dx, \quad n \geq 2. \quad (6.32)$$

We omit case $n = 1$, because estimates for

$$E_F(R_1 - R_0) = E_F R_1 - \mu_F$$

were already presented. Properties of functions

$$\begin{aligned} \varphi_n(x) &= f_n(x) - f_{n-1}(x) \\ &= \frac{[-\ln(1-x)]^n}{n!} - \frac{[-\ln(1-x)]^{n-1}}{(n-1)!} \\ &= f_{n-1}(x)\left[\frac{-\ln(1-x)}{n} - 1\right] \quad n \geq 2, \quad (6.33) \end{aligned}$$

are crucial for determining its projection onto various convex cones. We easily check that each φ_n integrates to 0, starts from the origin, decreases to

$$\varphi(1 - e^{-n+1}) = -\frac{(n-1)^{n-1}}{n!},$$

and increases to $+\infty$ at 1 passing through the horizontal axis at $1 - e^{-n}$. Its antiderivative Φ_n needed for the Moriguti [58] projection, is therefore concave decreasing, convex decreasing, and convex increasing in $[0, 1 - e^{-n+1}]$, $[1 - e^{-n+1}, 1 - e^{-n+1}]$, and $[1 - e^{-n}, 1]$, respectively. This vanishes at 0 and 1, and is negative in the middle. Thus we deduce that its greatest convex minorant $\bar{\Phi}_n$ is linear in $[0, \alpha_*]$ for some $\alpha_* \in [1 - e^{-n+1}, 1 - e^{-n+1}]$, that is determined by equation

$$x\varphi_n(x) = \Phi_n(x) = F_n(x) - F_{n-1}(x) = f_n(x) \qquad (6.34)$$

(cf. (6.4)), and coincides with Φ_n elsewhere. Finally,

$$P^0 \varphi_n(x) = \bar{\Phi}'_n(x) = \varphi_n(\max\{\alpha_*, x\}).$$

Theorem 37 (general distributions) *For $n \geq 2$ we have*

$$\frac{E_F(R_n - R_{n-1})}{\sigma_F} \leq \Delta = \Delta(n), \qquad (6.35)$$

where

$$\Delta^2(n) = \binom{2n-2}{n-1}(1 - \alpha_*) \left[\sum_{j=0}^{2n-1} \frac{(n\alpha_*)^j}{j!} + \left(1 - \frac{1}{n}\right) \frac{(n\alpha_*)^{2n-1}}{(2n-1)!} \right] \qquad (6.36)$$

for unique $\alpha_* \in (1 - e^{-n+1}, 1 - e^{-n+1})$ satisfying equation

$$-\ln(1 - x) = nx. \qquad (6.37)$$

Equality in (6.35) holds for

$$F(x) = \varphi_n^{-1}\left(\Delta \frac{x - \mu}{\sigma}\right) \mathbf{1}_{[\mu - \sigma\varphi_n(\alpha_*)/\Delta, \infty)}(x). \qquad (6.38)$$

Formula (6.37) is a reduction of (6.34), and $\|\bar{\Phi}'_n\|^2$ is rewritten as (6.36). We can also show that

$$\Delta(n) \sim \binom{2n-2}{n-1}^{1/2} \sim \frac{2^{n-1}}{(n\pi)^{1/4}},$$

which is the rate of increase of the extreme expectation of the $(n-1)$st record value (cf. (6.2)). Distribution (6.38) has jump α_* and a density with infinite support right to the jump point. Observe that the contribution of the smooth component $1 - \alpha_* < e^{-n+1}$ is very small.

We are able to establish similar bounds for populations with monotone density and failure rate functions, once we apply the following observation.

Lemma 27 *Decreasing-increasing functions $\varphi_n(x)$ and $\psi_n(x) = \varphi_n(1 - e^{-x})$, $n \geq 2$, are convex on their intervals of increase.*

PROOF. For
$$\psi_n(x) = \frac{x^n}{n!} - \frac{x^{n-1}}{(n-1)!}$$
that increases for $x \geq n - 1$, we simply verify the claim by repeated differentiation. Since φ_n is the superposition $\psi_n(x)$ with the increasing convex function $V^{-1}(x) = -\ln(1 - x)$, we have
$$\varphi_n''(x) = \psi_n'' V^{-1}(x)[V^{-1}(x)]^2 + \psi_n' V^{-1}(x)(V^{-1})''(x) > 0$$
for $x > 1 - e^{-n+1}$, which is the minimum point of φ_n. The relation is implied by the positivity of all the above components of $\varphi_n''(x)$. ∎

Proposition 28 (decreasing density and failure rate) *For $F \succeq_c U$ bound (6.35) is the best possible.*

The conclusion for $F \succeq_c V$ follows from $V \succ_c U$. The statement of the proposition can be proved by checking that (6.38) has a decreasing density, which is equivalent to the convexity of $\varphi_n(\max\{\alpha_*, x\})$. Since $\alpha_* > 1 - e^{-n+1}$, this immediately follows from Lemma 27.

Proposition 29 (increasing density and failure rate) *For $F \preceq_c U$*
$$\frac{E_F(R_n - R_{n-1})}{\sigma_F} \leq \frac{\sqrt{3}}{2^n}, \tag{6.39}$$
which is the equality for the uniform distribution on an arbitrary interval of length $2\sqrt{3}\sigma$.

For every $F \preceq_c V$
$$\frac{E_F(R_n - R_{n-1})}{\sigma_F} \leq 1, \tag{6.40}$$
which becomes the equality for the exponential distribution with scale σ.

Both assertions are deduced by means of analogous arguments. The crucial steps of the proofs consist in showing that projections of φ_n and ψ_n onto the cones of nondecreasing concave functions are linear. By Lemma 2, neither φ_n (ψ_n) nor the projection can majorize the other. Since both φ_n and ψ_n are first decreasing and then strictly convex increasing, every nondecreasing concave function may cross either of them at two points at most. Taking linear functions that pass through the crossing points improves the approximations of φ_n and ψ_n. Standard arguments of the projection method lead us to the final conclusions.

6.5 Open Problems

1. An annoying still unsolved problem is to find the bounds on the nth values of the kth records in symmetric populations for general n and k. The trouble here lies in describing variability of the function

$$\begin{aligned} s_n^{(k)}(x) &= f_n^{(k)}(x) - f_n^{(k)}(1-x) \\ &= \frac{k^{n+1}}{n!}\{[-\ln(1-x)]^n(1-x)^{k-1} - [-\ln x]^n x^{k-1}\} \end{aligned}$$

for $1/2 < x < 1$. This is necessary for application of the greatest convex minorant construction. Furthermore, we could ask for respective stricter bounds for symmetric unimodal distributions. Note that these are unknown in the case of first records as well.

2. What are the mean-variance bounds for kth records of i.i.d. samples with decreasing density and failure rate? For distributions with increasing density and failure rate neither second moment nor mean-variance bounds are known. A more general problem consists in establishing analogous results for distributions preceding and succeeding a given one in the convex order.

3. What are the respective evaluations for the record values coming from populations with distributions determined by the star order relations with a fixed one?

4. Results of Section 6.4 should be completed by considering increments of kth records $R_n^{(k)} - R_{n-1}^{(k)}$ in various classes of parent distributions. It is of interest to compare kth record increments for different k.

5. We have practically no evaluations for the lower records. Generally, they cannot be concluded from analogous results determined for the upper records.

6. Establishing bounds on $E_F(R_n - X_{n:n})$ in various families of distributions we evaluate the rate of increase of the gap between the elements of nondecreasing sequences of sample maxima and their strictly increasing subsequence. Analyzing $E_F(R_n^{(k)} - X_{n+1-k:n})$, we derive similar evaluations for the kth largest order statistics.

7
Predictions of Order and Record Statistics

In this chapter we evaluate expected increments of future order and record statistics in the i.i.d. samples under conditions that some previous values are known. These are important for predicting prospective failures in reliability systems and shock models on the grounds of former data. The results are based on representations of conditional expectations of order statistics and records in terms of unconditional expectations of other ones presented in Section 2.2, and evaluations of the latter derived in Sections 4.1, 4.2, 6.2, and 6.3. These provide bounds in terms of conditional second moments

$$m^2_{F^0_{|y}} = \mathrm{E}_F((X-y)^2|X>y) = \frac{\int_{F^{-1}(y)}^1 [F^{-1}(x) - y]^2\, dx}{1 - F(y)}, \qquad (7.1)$$

where y is the value of a previous observation. We have

$$\begin{aligned}
m^2_{F^0_{|y}} &\leq \frac{\int_0^1 [F^{-1}(x) - y]^2\, dx}{1 - F(y)} \\
&= \frac{\mathrm{E}_F(X-y)^2}{1 - F(y)} \\
&= \frac{\sigma_F^2 + (\mu_F - y)^2}{1 - F(y)} \\
&= M^2_{F^0_{|y}} = M, \qquad (7.2)
\end{aligned}$$

say. The equality in (7.2) holds if

$$F^{-1}(x) = y, \qquad x \in [0, F^{-1}(y)];$$

that is, F has a jump at the left endpoint y of its support. It occurs that the optimal constants for bounds expressed in terms of $m_{F^0_{|y}}$ are identical with those measured in larger units $M_{F^0_{|y}}$. These have a more intuitive meaning depending on moments and value of the parent distribution at a fixed observation point, and allow us to determine distributions attaining bounds uniquely up to the three mentioned parameters. The former are independent of the distribution at points preceding y, and so admit an ambiguity in the description of extremal distributions. Therefore we choose representations of bounds in terms of $M_{F^0_{|y}}$-units. Results for general distributions, and ones following the uniform and exponential distributions in the convex order, are stated in Sections 7.1 and 7.2, respectively. Formal proofs will be published in Rychlik [91].

7.1 General Distributions

We first consider predictions of order statistics. Our aim is to estimate expected increments of order statistics $X_{j:n} - X_{i:n}$ for some $1 \leq i < j \leq n$ when $X_{i:n} = y$ is known. By (2.41),

$$E_F(X_{j:n} - X_{i:n}|X_{i:n} = y) = E_{F_{|y}} X_{j-i:n-i} - y = E_{F^0_{|y}} X_{j-i:n-i}, \quad (7.3)$$

where F is a continuous distribution function with finite mean μ_F and variance σ_F^2, and

$$F^0_{|y}(x) = F_{|y}(x+y) = \frac{F(x+y) - F(y)}{1 - F(y)}, \quad x \geq 0, \quad (7.4)$$

is the distribution function of $X - y$ under the condition that $X > y$. Set $F(y) = p$. Below we describe the maximal values of (7.3) for all distributions with given μ_F, σ_F, and $F(y) = p$ measured in scale units

$$M_{F^0_{|y}} = \left[\frac{\sigma_F^2 + (\mu_F - y)^2}{1 - p} \right]^{1/2}.$$

Note first that quantile functions of (7.4)

$$(F^0_{|y})^{-1}(x) = F^{-1}(p + (1-p)x) - y \quad (7.5)$$

form the convex cone $\mathcal{C}^+$ of nondecreasing elements of $L^2([0,1), dx)$ starting from 0. Combining (7.3) and (7.4) with (2.23), and further applying the projection method, the Schwarz inequality, and (7.2), we obtain

$$\begin{aligned}
\mathrm{E}_F(X_{j:n} - X_{i:n}|X_{i:n} = y) &= \int_0^1 [F^{-1}(p + (1-p)x) - y] f_{j-i:n-i}(x)\, dx \\
&\leq \int_0^1 [F^{-1}(p + (1-p)x) - y] \bar{f}_{j-i:n-i}(x)\, dx \\
&\leq \|\bar{f}_{j-i:n-i}\| m_{F^0_{|y}} \\
&\leq \|\bar{f}_{j-i:n-i}\| M_{F^0_{|y}}, \quad (7.6)
\end{aligned}$$

where $\bar{f}_{j-i:n-i}$ is the projection of $f_{j-i:n-i}$ onto the family of nondecreasing functions derived by means of the greatest convex minorant method. The equality in (7.6) holds when $F^{-1}(p+(1-p)x) - y$ is proportional to $\bar{f}_{j-i:n-i}$, and equal to zero for arguments $z = p + (1-p)x \in [p,1)$, and $z \in [0,p)$, respectively. Note that for $j = n$ constructing the projection is redundant, because $f_{n-i:n-i}$ is actually increasing. If $j = i+1$, the projection is constant amounting to 1. Explicit representations of bounds are described in Theorem 38.

Theorem 38 (order statistics, general distributions) *Sharp bound*

$$\frac{\mathrm{E}_F(X_{i+1:n} - X_{i:n}|X_{i:n} = y)}{M_{F^0_{|y}}} \leq 1 \quad (7.7)$$

is attained by the two-point distribution

$$\mathrm{P}(X = y) = p = 1 - \mathrm{P}(X = y + M_{F^0_{|y}}). \quad (7.8)$$

For $i + 2 \leq j \leq n - 1$, we have

$$\frac{\mathrm{E}_F(X_{j:n} - X_{i:n}|X_{i:n} = y)}{M_{F^0_{|y}}} \leq \|\bar{f}_{j-i:n-i}\|, \quad (7.9)$$

where the optimal constant $\|\bar{f}_{j-i:n-i}\|$ is defined in (4.10) with parameter $\alpha_* = \alpha_*(j-i, n-i)$ determined by Equation (4.7). The equality in (7.9) holds for

$$F(x) = \begin{cases} 0, & \text{if } \frac{x-y}{M} < 0, \\ p + (1-p) f^{-1}_{j-i:n-i}(\|\bar{f}_{j-i:n-i}\| \frac{x-y}{M}), & \text{if } 0 \leq \frac{x-y}{M} < \frac{f_{j-i:n-i}(\frac{\alpha_*-p}{1-p})}{\|\bar{f}_{j-i:n-i}\|}, \\ 1, & \text{if } \frac{x-y}{M} \geq \frac{f_{j-i:n-i}(\frac{\alpha_*-p}{1-p})}{\|\bar{f}_{j-i:n-i}\|}. \end{cases}$$

$$(7.10)$$

Finally,

$$\frac{\mathrm{E}_F(X_{n:n} - X_{i:n}|X_{i:n} = y)}{M_{F^0_{|y}}} \leq \frac{n-i}{(2n - 2i - 1)^{1/2}} \quad (7.11)$$

holds and becomes the equality for the combination of degenerate and power distributions

$$F(x)=\begin{cases} 0, & \text{if } \frac{x-y}{M} < 0, \\ p+(1-p)[\frac{1}{\sqrt{2n-2i-1}}\frac{x-y}{M}]^{1/(n-i-1)}, & \text{if } 0 \le \frac{x-y}{M} < \sqrt{2n-2i-1}, \\ 1, & \text{if } \frac{x-y}{M} \ge \sqrt{2n-2i-1}. \end{cases} \quad (7.12)$$

All the distribution functions (7.8), (7.10), and (7.12) have jumps at y. Also, (7.10) has an atom at the right end of its support, and (7.12) does not. We emphasize the fact that basic relation (7.3) and, accordingly, issuing inequalities (7.7), (7.9), and (7.11) hold true for continuous distributions F. They are the best possible, but attainable merely in the limit by continuous distributions tending to (7.8), (7.10), and (7.12), respectively, in the sense described earlier (see, e.g., the last two paragraphs of Section 2.3). The same reservations concern the other results of this chapter.

We now proceed to records and make use of identity

$$E_F(R_n^{(k)} - R_m^{(k)} | R_m^{(k)} = y) = E_{F_{|y}^0} R_{n-m-1}^{(k)}, \quad n > m, \quad (7.13)$$

concluded from (2.44). Since for $n = m+1$ we have

$$E_F(R_{m+1}^{(k)} - R_m^{(k)} | R_m^{(k)} = y) = E_{F_{|y}^0} X_{1 \cdot k},$$

the evaluations of the first statement of Theorem 38 (see (7.7) and (7.8)) concerning the conditional expectations of spacings apply to the differences of consecutive record values as well. We omit the case in our further considerations. Mimicking arguments applied in (7.6), we obtain

$$\begin{aligned} E_F(R_n^{(k)} - R_m^{(k)} | R_m^{(k)} = y) &= \int_0^1 [F^{-1}(p+(1-p)x) - y] f_{n-m-1}^{(k)}(x)\, dx \\ &\le \int_0^1 [F^{-1}(p+(1-p)x) - y] \bar{f}_{n-m-1}^{(k)}(x)\, dx \\ &\le \|\bar{f}_{n-m-1}^{(k)}\| M_{F_{|y}^0}. \end{aligned} \quad (7.14)$$

For $k = 1$, we have $\bar{f}_{n-m-1}^{(k)} = f_{n-m-1}^{(k)}$ which are actually increasing. Otherwise the Moriguti [58] construction of nondecreasing approximations is necessary.

Theorem 39 (kth record values, general distributions) *For $k = 1$ (standard record values), we have*

$$\frac{E_F(R_n - R_m | R_m = y)}{M_{F_{|y}^0}} \le \left(\frac{2n - 2m - 2}{n - m - 1}\right)^{1/2}, \quad (7.15)$$

which becomes the equality for the mixture of Dirac and Weibull distributions

$$F(x) = \left\{ p + (1-p) \left[1 - \exp\left(-\left[\sqrt{(2n-2m-2)!} \frac{x-y}{M} \right]^{1/(n-m-1)} \right) \right] \right\} \mathbf{1}_{[0,\infty)} \left(\frac{x-y}{M} \right). \tag{7.16}$$

If $k \geq 2$, then

$$\frac{E_F(R_n^{(k)} - R_m^{(k)} | R_m^{(k)} = y)}{M_{F_{|y}^0}} \leq \|\bar{f}_{n-m-1}^{(k)}\|, \tag{7.17}$$

where the right-hand side is defined in (6.6) with $\alpha_* = \alpha_*(k, n-m-1)$ coming from (6.4). The equality in (7.17) holds for (7.10) with $f_{j-i:n-i}$ replaced by $f_{n-m-1}^{(k)}$.

7.2 Distributions with Decreasing Density and Failure Rate

Here we confine ourselves to restricted families of parent distributions defined by relation $F \succeq_c W$ with a given W. We first notice that $F \succeq_c W$ does not entail the same for $F_{|y}^0$ which appears in unconditional representations (7.3) and (7.13) of conditional expectations of the increments. This is shown by the following counterexample.

EXAMPLE 4. Set $F(x) = x^{1/3}$ and $W(x) = x^{1/2}$, $0 \leq x \leq 1$. Obviously $F^{-1}W(x) = x^{3/2}$ is convex, and so $F \succ_c W$. Take now

$$F_{|1/8}^0(x) = 2(x + 1/8)^{1/3} - 1, \quad 0 \leq x \leq 7/8,$$

with the quantile function

$$(F_{|1/8}^0)^{-1}(x) = \frac{1}{8}[(x+1)^3 - 1], \quad 0 \leq x \leq 1.$$

The composition

$$(F_{|1/8}^0)^{-1} W(x) = \frac{1}{8}[(x^{1/2} + 1)^3 - 1] = \frac{1}{8}(x^{3/2} + 3x + 3x^{1/2})$$

has the second derivative

$$[(F_{|1/8}^0)^{-1} W]''(x) = \frac{3}{32} x^{-3/2}(x - 1),$$

negative for $x \in [0, 1)$, which proves $F_{|1/8}^0 \prec_c W$. ∎

However, relation $F \succeq_c W$ is preserved if we compare $F_{|y}^0$ with similar modification $W_{|z}^0$ for properly chosen z such that $W(z) = F(y) = p$. Indeed, we have

$$(F_{|y}^0)^{-1} W_{|z}^0(x) = F^{-1} W(x+z) - y, \quad 0 \leq x < d_W - z, \quad (7.18)$$

whose convexity is implied by that of $F^{-1}W(x)$. Note that the reversed implication is false, because we can extend $F_{|y}^0$ and $W_{|z}^0$ on the left to y and z, respectively, in an arbitrary way. Compositions (7.18) form the convex cone

$$\mathcal{C}^+_{\succeq_c W_{|z}^0} \subset L^2\left([0, d_W - z), \frac{w(z+x)}{1 - W(z)} dx\right).$$

In particular, for uniform $W = U$, compositions (7.18) have the form

$$(F_{|y}^0)^{-1} U_{|p}^0(x) = F^{-1}(x+p) - y, \quad 0 \leq x < 1 - p,$$

and

$$(F_{|y}^0)^{-1} U_{|p}^0 \in \mathcal{C}^+_{\succeq_c U_{|p}^0} \subset L^2\left([0, 1-p), \frac{dx}{1-p}\right).$$

In the exponential case $W = V$, we have

$$(F_{|y}^0)^{-1} V_{|z}^0(x) = F^{-1}(1 - (1-p)e^{-x}) - y, \quad 0 \leq x < \infty,$$

and the identity

$$\mathcal{C}^+_{\succeq_c V_{|z}^0} = \mathcal{C}^+_{\succeq_c V} \subset L^2([0, \infty), e^{-x} dx)$$

holds for all positive z.

We are now in a position to formulate the main results of this section. We first focus on differences of consecutive order statistics and records. Projecting the respective functionals we obtain

$$\mathbb{E}_F(X_{i+1:n} - X_{i:n} | X_{i:n} = y) \leq \mu_{F_{|y}^0} \leq m_{F_{|y}^0} \leq M_{F_{|y}^0}, \quad (7.19)$$

$$\mathbb{E}_F(R_{m+1}^{(k)} - R_m^{(k)} | R_m^{(k)} = y) \leq \mu_{F_{|y}^0} \leq m_{F_{|y}^0} \leq M_{F_{|y}^0} \quad (7.20)$$

(cf. (7.1)). Equalities in (7.19) and (7.20) hold if $F^{-1}(x)$ is constant for $x \in [p, 1)$. Convexity and nondecrease of $F^{-1}W(x)$ impose the single value of the quantile function on the whole domain. The conditional expectations have a practical sense if y is the only support point of F, and they are equal to 0, but all the evaluations are meaningless then.

Some results of this section can be immediately concluded from the previous one. For instance, (7.12) is a decreasing density distribution function

7.2 Distributions with Decreasing Density and Failure Rate

which implies that (7.11) is the best possible bound for the restricted class. In the remaining cases $3 \leq i+2 \leq j < n$ we have

$$
\begin{aligned}
& E_F(X_{j:n} - X_{i:n}|X_{i:n} = y) \\
&= \int_0^{1-p} [F^{-1}(x+p) - y] f_{j-i:n-i}\left(\frac{x}{1-p}\right) \frac{dx}{1-p} \\
&\leq \int_0^{1-p} [F^{-1}(x+p) - y] P^+_{\succeq_c U} f_{j-i:n-i}\left(\frac{x}{1-p}\right) \frac{dx}{1-p} \\
&\leq \|P^+_{\succeq_c U} f_{j-i:n-i}\| m_{F^0_{|y}} \\
&\leq \|P^+_{\succeq_c U} f_{j-i:n-i}\| M_{F^0_{|y}}. \qquad (7.21)
\end{aligned}
$$

The equality is a consequence of (7.3), (7.5), and a change of variables. The first inequality follows from (2.2) and relation

$$
P^+_{\succeq_c U^0_{|p}} f_{j-i:n-i}(x) = P^+_{\succeq_c U} f_{j-i:n-i}\left(\frac{x}{1-p}\right),
$$

which is implied by the equivalence of claims

$$
g\left(\frac{\cdot}{1-p}\right) \in C^+_{\succeq_c U^0_{|p}} \iff g(\cdot) \in C^+_{\succeq_c U},
$$

and the identity

$$
\int_0^{1-p} \left[f_{j-i:n-i}\left(\frac{x}{1-p}\right) - g\left(\frac{x}{1-p}\right)\right]^2 \frac{dx}{1-p} = \int_0^1 [f_{j-i:n-i}(x) - g(x)]^2 dx.
$$

The second inequality in (7.21) follows from the Schwarz inequality combined with changes of variables, and (7.2) implies the last one. Equality in (7.21) is obtained if

$$
\begin{aligned}
F^{-1}(x+p) - y &= \frac{P^+_{\succeq_c U} f_{j-i:n-i}\left(\frac{x}{1-p}\right)}{\|P^+_{\succeq_c U} f_{j-i:n-i}\|} M_{F^0_{|y}}, \quad 0 \leq x < 1, \quad (7.22) \\
F^{-1}(x) &= y, \qquad\qquad\qquad\qquad\qquad\qquad 0 \leq x < p. \quad (7.23)
\end{aligned}
$$

Recalling Theorem 11, we can now state the following assertions.

Theorem 40 (order statistics, decreasing density) *If $i + 2 \leq j \leq \min\{(2n + i + 2)/3, n - 1\}$, then*

$$
\frac{E_F(X_{j:n} - X_{i:n}|X_{i:n} = y)}{M_{F^0_{|y}}} \leq \sqrt{3}\frac{j-i}{n+1-i}, \qquad (7.24)
$$

which becomes the equality for F being the convex combination of the Dirac measure at y and the uniform distribution on $[y, y + \sqrt{3}M]$ with respective coefficients p and $1 - p$.

If $(2n+i+2)/3 < j \leq n-1$, then

$$\frac{E_F(X_{j:n} - X_{i:n}|X_{i:n} = y)}{M_{F^0_{|y}}} \leq ||(f_{j-i:n-i})_{\alpha_*\beta_*}||, \qquad (7.25)$$

where the right-hand side is defined by (4.46) with $\alpha_* = \alpha_*(\beta_*)$ and $\beta_* = \beta_*(j-i, n-i)$ uniquely determined by (4.47) through (4.49). Inequality (7.25) becomes the equality for

$$F(x) = \begin{cases} 0, \\ p + (1-p)f^{-1}_{j-i:n-i}(||(f_{j-i:n-i})_{\alpha_*\beta_*}||\frac{x-y}{M}), \\ p + (1-p)\{\beta_* + \frac{1}{\alpha_*}[||(f_{j-i:n-i})_{\alpha_*\beta_*}||\frac{x-y}{M} - f_{j-i:n-i}(\beta_*)]\}, \\ 1, \end{cases} \qquad (7.26)$$

if
$$\begin{cases} \frac{x-y}{M} < 0, \\ 0 \leq \frac{x-y}{M} \leq \frac{f_{j-i:n-i}(\beta_*)}{||(f_{j-i:n-i})_{\alpha_*\beta_*}||}, \\ \frac{f_{j-i:n-i}(\beta_*)}{||(f_{j-i:n-i})_{\alpha_*\beta_*}||} \leq \frac{x-y}{M} \leq \frac{f_{j-i:n-i}(\beta_*)+\alpha_*(1-\beta_*)}{||(f_{j-i:n-i})_{\alpha_*\beta_*}||}, \\ \frac{x-y}{M} \geq \frac{f_{j-i:n-i}(\beta_*)+\alpha_*(1-\beta_*)}{||(f_{j-i:n-i})_{\alpha_*\beta_*}||}, \end{cases} \qquad (7.27)$$

respectively.

In the case $F \succeq_c V$, the projection method is essentially exploited for predicting sample maxima as well. By arguments similar to ones used in (7.21), for $i+2 \leq j \leq n$ we deduce that

$$E_F(X_{j:n} - X_{i:n}|X_{i:n} = y)$$
$$= \int_0^\infty [F^{-1}(p + (1-p)V(x)) - y]f_{j-i:n-i}V(x)v(x)\,dx$$
$$\leq \int_0^\infty [F^{-1}(p + (1-p)V(x)) - y]P^+_{\succeq_c V}f_{j-i:n-i}V(x)v(x)\,dx$$
$$\leq ||P^+_{\succeq_c V}f_{j-i:n-i}V||M_{F^0_{|y}}, \qquad (7.28)$$

with the equality valid under conditions

$$F^{-1}(p+(1-p)V(x)) - y = \frac{P^+_{\succeq_c V}f_{j-i:n-i}V(x)}{||P^+_{\succeq_c V}f_{j-i:n-i}V||}M_{F^0_{|y}}, \qquad x \geq 0,$$
$$F^{-1}(x) = y, \qquad 0 \leq x < p$$

(cf. (7.22) and (7.23)). Finally, we refer to Theorem 12.

Theorem 41 (order statistics, decreasing failure rate) *Under notation (4.51), if $\mu_{V_{j-i:n-i}} \leq 2$, then*

$$\frac{E_F(X_{j:n} - X_{i:n}|X_{i:n} = y)}{M_{F^0_{|y}}} \leq \frac{\mu_{V_{j-i:n-i}}}{\sqrt{2}}. \qquad (7.29)$$

7.2 Distributions with Decreasing Density and Failure Rate

This becomes the equality if F is the mixture of an atom at y and the exponential distribution with location y and scale $M/\sqrt{2}$, with respective coefficients p and $1-p$.

Otherwise

$$\frac{\mathrm{E}_F(X_{j:n} - X_{i:n}|X_{i:n} = y)}{M_{F^0_{|y}}} \leq \|(f_{j-i:n-i}V)_{\alpha_*}V^{-1}(\gamma_*)\|, \qquad (7.30)$$

with the norm and its parameters $\alpha_* = \alpha_*(\gamma_*)$ and $\gamma_* = \gamma_*(j-i, n-i)$ defined in (4.54) through (4.57). Equality holds in (7.30) for

$$F(x) = \begin{cases} 0, \\ p + (1-p)f_{j-i:n-i}^{-1}(\|(f_{j-i:n-i}V)_{\alpha_*}V^{-1}(\gamma_*)\|\frac{x-y}{M}), \\ p + (1-p)\{\gamma_* + (1-\gamma_*) \\ \quad \times V(\frac{1}{\alpha_*}[\|(f_{j-i:n-i}V)_{\alpha_*}V^{-1}(\gamma_*)\|\frac{x-y}{M} - f_{j-i:n-i}(\gamma_*)])\}, \end{cases} \qquad (7.31)$$

$$if \begin{cases} \frac{x-y}{M} < 0, \\ 0 \leq \frac{x-y}{M} \leq \frac{f_{j-i:n-i}(\gamma_*)}{\|(f_{j-i:n-i}V)_{\alpha_*}V^{-1}(\gamma_*)\|}, \\ \frac{f_{j-i:n-i}(\gamma_*)}{\|(f_{j-i:n-i}V)_{\alpha_*}V^{-1}(\gamma_*)\|} \leq \frac{x-y}{M}, \end{cases} \qquad (7.32)$$

respectively.

Note that (7.16) has decreasing density and failure rate. This means that general bounds (7.15) are the optimal ones for the predictions of the values of the first records from populations with decreasing density and failure rate. Therefore it remains to study differences of nonsuccessive values of kth records for $k \geq 2$. In order to obtain desired conclusions, we are reduced to replacing $f_{j-i:n-i}(x)$ by $f^{(k)}_{n-m-1}(x)$ in (7.21) and (7.28), and references to Theorems 35 and 36.

Theorem 42 (kth records, decreasing density) *If $(1+1/k)^{n-m} \leq 3$, we have*

$$\frac{\mathrm{E}_F(R^{(k)}_n - R^{(k)}_m | R^{(k)}_m = y)}{M_{F^0_{|y}}} \leq \sqrt{3}\left[1 - \left(\frac{k}{k+1}\right)^{n-m}\right], \qquad (7.33)$$

which is the equality if F is the mixture of the uniform distribution on $[y, y+\sqrt{3}M]$ and an atom at y.

In the opposite case, we have

$$\frac{\mathrm{E}_F(R^{(k)}_n - R^{(k)}_m | R^{(k)}_m = y)}{M_{F^0_{|y}}} \leq \|(f^{(k)}_{n-m-1})_{\alpha_*\beta_*}\|, \qquad (7.34)$$

defined in (6.20) with parameters $\alpha_* = \alpha_*(\beta_*)$ and $\beta_* = \beta_*(k, n-m-1)$ defined in (6.21) to (6.23). Bound (7.34) is attained by the distribution function given in (7.26) and (7.27) with $f_{j-i:n-i}$ replaced by $f^{(k)}_{n-m-1}$.

Theorem 43 (kth records, decreasing failure rate) If $m + 2 \leq n \leq m + 2k$, then

$$\frac{\mathrm{E}_F(R_n^{(k)} - R_m^{(k)}|R_m^{(k)} = y)}{M_{F_{|y}^0}} \leq \frac{n-m}{\sqrt{2k}}. \qquad (7.35)$$

This is the equality for the combination of the exponential distribution with location parameter y and scale $M/\sqrt{2}$, and an atom at y.

Otherwise

$$\frac{\mathrm{E}_F(R_n^{(k)} - R_m^{(k)}|R_m^{(k)} = y)}{M_{F_{|y}^0}} \leq \|(f_{n-m-1}^{(k)} V)_{\alpha_*} V^{-1}(\gamma_*)\|, \qquad (7.36)$$

where the bound is defined by (6.27), and the parameters $\alpha_* = \alpha_*(\gamma_*)$ and $\gamma_* = \gamma_*(k, n-m-1)$ are determined from (6.28) to (6.30). Inequality (7.36) becomes the equality for F defined in (7.31) and (7.32) with $f_{j-i:n-i}$ replaced by $f_{n-m-1}^{(k)}$.

Distribution functions (7.26), (7.31), and their modifications described in Theorems 42 and 43 have unbounded derivatives in the right neighborhoods of the left endpoints of their supports. The shape conditions entail unique extensions of them to the left: only a jump at y and no mass of the left are admitted. This means that the analogous bounds expressed in $m_{F_{|y}^0}$-units are attained by the same distributions only. This is not so in the remaining cases. It is worth noticing that bounds (7.24) and (7.33) as well as (7.29) and (7.35) are attained by the same elements of families with decreasing density and failure rate, respectively.

7.3 Open Problems

1. What are the sharp bounds on conditional expectations of increments of order and record statistics from populations with increasing density and failure rate? More generally, we ask about respective evaluations for the families of distributions determined by the convex order relations with a fixed abstract distribution function W.

2. Solve the analogous problem for the distributions defined by the star order relations.

3. Applying (2.42) and the projection method makes it possible to recover the conditional of past failures. Similar results for previous record values can be established by exploiting the dependence structure of records.

7.3 Open Problems

4. One can try to retrieve missing values of order statistics using (2.43). However, the problem is that the quantile functions of doubly truncated distributions $F_{|y}^{|z}$ do not form a convex cone, and one cannot simply use our projection method here.

5. In contrast to the previous results, all the bounds presented in Chapter 7 are expressed in terms of sophisticated scale units

$$M_{F_{|y}^0} = \left[\frac{\sigma_F^2 + (\mu_F - y)^2}{1 - F(y)}\right]^{1/2}.$$

Is it possible to replace them by simpler and more intuitive ones? Certainly, the problem also surpasses the range of direct applications of projections onto convex cones.

8
Further Research Directions

In the monograph we mostly focused on the optimal upper bounds, but certainly the lower ones are needed for evaluating the actual ranges of the functionals over given classes of distributions. Generally, the best upper and lower bounds are not symmetric about zero, but the latter can also be derived by means of our projection method. To this end we should analyze the negatives of the functionals under study. Only the functional corresponding to the order statistics and L-statistics (see (2.27) and (2.25)) in the dependent case needs a more subtle transformation. Changing the signs of coefficients c_j, $1 \leq j \leq n$, in (2.26) results in constructing a functional $g_{-\mathbf{c}}$ different from $-g_{\mathbf{c}}$. Also, one should realize that generally projecting a functional and its negative are different problems that should be solved separately by use of specific arguments. With few exceptions they cannot be derived one from the other in a simple way.

Here we confined ourselves to statistical functionals defined for finite samples, but various notions of asymptotic statistics are represented in that form. As an example we mention the limits

$$T_h(F^{-1}) = \int_0^1 F^{-1}(x) h(x)\, dx$$

of sequences of L-statistics

$$\sum_{j=1}^n c_{j,n} X_{j:n} = \sum_{j=1}^n \left[h\left(\frac{j}{n}\right) - h\left(\frac{j-1}{n}\right) \right] X_{j:n}$$

with coefficients determined by a smooth weight function h. From a practical point of view, we are interested in uniform bias evaluation for finite sample estimates of asymptotic quantities in various classes of distributions. In our example, it is to analyze the differences

$$E_F \sum_{j=1}^{n} c_{j,n} X_{j:n} - \int_0^1 F^{-1}(x) h(x)\, dx$$

$$= \int_0^1 F^{-1}(x) \left\{ \left[h\left(\frac{j}{n}\right) - \left(\frac{j-1}{n}\right) \right] f_{j:n}(x) - h(x) \right\} dx.$$

An interesting topic of investigation is analyzing functionals of generalized order statistics extensively presented in Kamps [42]. The *generalized order statistics* $X(j, n, m, k)$, $j = 1, \ldots n$, based on distribution function F with real parameters m, k satisfying

$$\eta_j = k + (n-j)(m+1) \geq 1, \quad 1 \leq j \leq n,$$

have expectations

$$E_F X(j,n,m,k) = \int_0^1 F^{-1}(x) \frac{\prod_{i=1}^{j} \eta_j}{(j-1)!(m+1)} (1-x)^{\eta_j - 1} [1 - (1-x)^{m+1}]^{j-1} dx$$

for $m \neq 1$, and

$$E_F X(j,n,-1,k) = \int_0^1 F^{-1}(x) \frac{\prod_{i=1}^{j} \eta_j}{(j-1)!} (1-x)^{\eta_j - 1} [-\ln(1-x)]^{j-1} dx.$$

If $m = 0$ and $k = 1$, then $X(j,n,m,k)$ reduce to the standard order statistics $X_{j:n}$, $1 \leq j \leq n$, of the independent sample with the common distribution function F. If F is absolutely continuous, k is a positive integer, and $m = -1$, then $X(j,n,m,k)$ do not depend on parameter n, and coincide with values of kth records $R_j^{(k)}$. For other choices of parameters, the generalized order statistics represent observations of some censoring and truncation schemes and complex reliability and shock models, for example sequential order statistics and Pfeifer's record model in which the failure probabilities of surviving elements change at the failure moments of the other ones.

As an example of other classes of distributions that admit our projection approach are the distributions determined by the relations of superadditive order with a given W. We say that F succeeds W in the *superadditive order* and write $F \succeq_+ W$ if $F^{-1}W$ is *superadditive*; that is,

$$F^{-1}W(x+y) \geq F^{-1}W(x) + F^{-1}W(y) \tag{8.1}$$

for positive x and y. The reversed inequality defines $F \preceq_+ W$. The superadditive order implies the star one and so describes larger classes of distributions. In fact, relation (8.1) is applied for the life distributions and should be satisfied for all $0 = a_W \leq x + y < d_W$, but it can be modified by subtracting $F^{-1}W(a_W) = F^{-1}(0)$ from all the terms in (8.1) when $a_W \neq 0$. Especially for $W = V$, composition

$$V^{-1}F(x) = -\ln[1 - F(x)]$$

defines the hazard function of F, and (8.1) describes the new worse than used (NWU) distributions. The interpretation of the subadditivity of the hazard function is that surviving time $x + y$ by a single device is more likely than under replacement by a new one in meantime x. The reversed relation defines the new better than used (NBU) distributions. Since mean-variance bounds presented here for $F \preceq_* W$ cannot be improved in general populations, the same holds for the larger classes $F \preceq_+ W$, and the NBU distributions in particular. However, a difficult problem is to determine bounds for $F \succeq_+ W$, because the superadditivity does not allow a natural graphical interpretation. Verifying the property of a function at fixed $x+y$, we should study its values at all pairs x and y. Since

$$g_{j:n}(x) = \frac{n}{n+1-j} \mathbf{1}_{[(j-1)/(n,1)]}(x)$$

is superadditive for $(j - 1/n \geq 1/2$, we have the conclusion of Corollary 1 for the subadditive distribution functions $F \succeq_+ U$ then. We conjecture that the bound on quantiles and order statistics of dependent samples of the NWU populations are attained by Poisson distributions.

The method based on the greatest convex minorants provides the best L^2-approximations of general functions by monotone ones. This was fruitfully exploited in determining sharp mean-variance and second moment bounds on various statistical functionals in general and symmetric populations. For other families of distributions, the projection heavily depends on properties of functions being projected. The reason is that there are no known procedures which allow us to determine projections of general functions onto corresponding convex cones of quantile functions and their modifications. What we strongly need here is, for example a general method of projecting onto the family of convex functions. It is not clear here whether one should try to generalize the Moriguti [58] method of convex minorants for this problem, or develop quite a different approach.

We obtain usually nonsharp bounds on the statistical functionals defined in Section 2.2 in terms of

$$\|F^{-1}\|_p = \left[\int_0^1 |F^{-1}(x)|^p \, dx \right]^{1/p}$$

and $\|F^{-1} - \mu_F\|_p$, $p > 1$, by using the Hölder inequality instead of the Schwarz one. These are scale parameters expressed in terms of pth roots

8. Further Research Directions

of the raw and central absolute moments, respectively, of order p. As we mentioned in the Introduction, general L^p-projections onto convex cones do not allow characterizations similar to (2.2) and (2.3) and, accordingly, do not give us optimal bounds in the scale units generated by pth moments. However, the L^2-projections onto the monotone functions can be modified so that we derive sharp evaluations in terms of the above scale parameters. Indeed, evaluating a nonzero statistical functional over the class of general quantile functions by means of (2.2) and the Hölder inequality, we have

$$\begin{aligned} T_h(F^{-1}) &= \int_0^1 F^{-1}(x) h(x)\, dx \\ &\leq \int_0^1 F^{-1}(x) \bar{h}(x)\, dx \\ &\leq \|\bar{h}\|_q \|F^{-1}\|_p \end{aligned} \qquad (8.2)$$

with $q = p/(p-1)$ and $\bar{h} = P^\nearrow h$ defined by means of the greatest convex minorant construction (see Example 3, Section 2.1). The Hölder inequality becomes the equality if either F^{-1} is zero or

$$F^{-1}(x) = \alpha \chi_p \bar{h}(x) = \alpha |\bar{h}(x)|^{q/p} \mathrm{sgn}(\bar{h}(x)) \qquad (8.3)$$

for some positive α. In particular, the equality holds for

$$F^{-1}(x) = \frac{\|F^{-1}\|_p}{\|\chi_p \bar{h}\|_p} \chi_p \bar{h}(x), \qquad (8.4)$$

which is actually a quantile function with a desired pth moment, because each χ_p, $p > 1$, is strictly increasing. Since the intervals on which (8.4) and the original $\bar{h}$ are constant coincide, the former satisfies the conditions of Lemma 3 for attaining equality in the second line of (8.2). Summing up, we have

$$\begin{aligned} T_h\left(\frac{\|F^{-1}\|_p}{\|\chi_p \bar{h}\|_p} \chi_p \bar{h}\right) &= \frac{\|F^{-1}\|_p}{\|\chi_p \bar{h}\|_p} \int_0^1 \chi_p \bar{h}(x) \bar{h}(x)\, dx \\ &= \|\bar{h}\|_q \|F^{-1}\|_p \\ &= \|\bar{h}\|_q \left\| \frac{\|F^{-1}\|_p}{\|\chi_p \bar{h}\|_p} \chi_p \bar{h} \right\|_p, \end{aligned}$$

which proves that (8.4) attains the equality in (8.2) that is actually sharp. Obvious modifications lead to sharp bounds for general and symmetric distributions in terms of $\|F^{-1} - \mu_F\|_p$. An important question is now whether L^2-projections onto convex cones of functions that obey conditions more stringent than monotonicity only admit modifications that provide sharp bounds in terms of general pth norms.

8. Further Research Directions 161

Finally, it is of interest if projections onto convex sets that are not necessarily convex cones provide meaningful evaluations of functionals. If so, we could significantly extend the class of functionals for which respective evaluations hold by adding ones that act directly on the distribution functions, densities, and other characteristics of distributions.

References

[1] Ahsanullah, M. (1995), *Record Statistics*, Nova Sci., Commack, NY.

[2] Arnold, B.C. (1980), Distribution-free bounds on the mean of the maximum of a dependent sample, *SIAM J. Appl. Math.* 38, 163–167.

[3] Arnold, B.C. (1985), p-Norm bounds on the expectation of the maximum of possibly dependent sample, *J. Multivar. Anal.* 17, 316–332.

[4] Arnold, B.C. (1988), Bounds on the expected maximum, *Commun. Statist. Theor. Meth.* 17, 2135–2150.

[5] Arnold, B. C. and N. Balakrishnan (1989), *Relations, Bounds and Approximations for Order Statistics*, Lecture Notes in Statistics, Vol. 53, Springer-Verlag, New York.

[6] Arnold, B.C. and R.A. Groeneveld (1974), Bounds for deviations between sample population statistics, *Biometrika* 61, 387–389.

[7] Arnold, B.C., N. Balakrishnan, and H.N. Nagaraja (1992), *A First Course in Order Statistics*, Wiley, New York.

[8] Arnold, B.C., N. Balakrishnan, and H.N. Nagaraja (1998), *Records*, Wiley, New York.

[9] Balakrishnan, A. V. (1981), *Applied Functional Analysis*, 2nd ed., Springer-Verlag, New York.

[10] Balakrishnan, N. (1993), A simple application of binomial-negative binomial relationship in the derivation of sharp bounds for moments of order statistics based on greatest convex minorants, *Statist. Probab. Lett.* 18, 301–305.

[11] Balakrishnan, N. and A.C. Cohen (1991), *Order Statistics and Inference: Estimation Methods*, Academic, Boston.

[12] Balakrishnan, N. and C.R. Rao (eds.) (1998), *Order Statistics: Theory & Methods*, Handbook of Statistics, Vol. 16, North Holland, Amsterdam.

[13] Balakrishnan, N. and C.R. Rao (eds.) (1998), *Order Statistics: Applications*, Handbook of Statistics, Vol. 17, North Holland, Amsterdam.

[14] Balakrishnan, N., C. Charalambides, and N. Papadatos (2001), Bounds on expectation of order statistics from a finite population, with an insight into Hartley–David–Gumbel, Samuelson–Scott, Arnold–Groeneveld and some other bounds, submitted for publication.

[15] Barlow, R.E. and F. Proschan (1966), Inequalities for linear combinations of order statistics from restricted families, *Ann. Math. Statist.* 37, 1574–1591.

[16] Biondini, R. and M.M. Siddiqui (1975), Record values in Markov sequences, in: *Statistical Inference and Related Topics*, Vol. 2 (M.L. Puri, ed.), Academic, New York, 291–352.

[17] Blom, G. (1958), *Statistical Estimates and Transformed Beta-Variables*, Almqvist and Wiksells, Uppsala.

[18] Boyd, A.V. (1971), Bound for order statistics, *Publ. of the Electrotechnical Faculty of Belgrade Univ., Math. Phys. Series* 365, 31–32.

[19] Caraux, G. and O. Gascuel (1992), Bounds on distribution functions of order statistics for dependent variates, *Statist. Probab. Lett.* 14, 103–105.

[20] Chandler, K.N. (1952), The distribution and frequency of record values, *J. Roy. Statist. Soc., Ser. B* 14, 220–228.

[21] Cheng, C. (1995), The Bernstein polynomial estimator of a smooth quantile function, *Statist. Probab. Lett.* 24, 321–330.

[22] David, H.A. (1981), *Order Statistics*, 2nd ed., Wiley, New York.

[23] David, H.A. (1988), General bounds and inequalities in order statistics, *Commun. Statist. Theor. Meth.* 17, 2119–2134.

[24] David, H.A., H.O. Hartley, and E.S. Pearson (1954), The distribution of the ratio, in a single normal sample, of range to standard deviation, *Biometrika* 41, 482–493.

[25] Dharmadhikari, S. and K. Joag-dev (1988), *Unimodality, Convexity, and Applications*, Academic, New York.

[26] Dziubdziela, W. and B. Kopociński (1976), Limiting properties of the kth record values, *Zastos. Mat.* 15, 187–190.

[27] Fahmy, S. and F. Proschan (1981), Bounds on differences of order statistics, *Amer. Statist.* 35, 46–47.

[28] Feldman, D. and H.G. Tucker (1966), Estimation of non-unique quantiles, *Ann. Math. Statist.* 37, 451-457.

[29] Franco, M. and J.M. Ruiz (1996), On characterization of continuous distributions by conditional expectation of record values, *Sankhyā A* 58, 135–141.

[30] Franco, M. and J.M. Ruiz (1999), Characterization based on conditional expectations of adjacent order statistics: A unified approach, *Proc. Amer. Math. Soc.* 127, 861–874.

[31] Gajek, L. and A. Okolewski (2001), Projection method for moment bounds on record statistics from restricted families, submitted for publication.

[32] Gajek, L. and T. Rychlik (1996), Projection method for moment bounds on order statistics from restricted families. I. Dependent case, *J. Multivar. Anal.* 57, 156–174.

[33] Gajek, L. and T. Rychlik (1998), Projection method for moment bounds on order statistics from restricted families. II. Independent case, *J. Multivar. Anal.* 64, 156–182.

[34] Gascuel, O. and G. Caraux (1992), Bounds on expectations of order statistics via extremal dependences, *Statist. Probab. Lett.* 15, 143–148.

[35] Grudzień, Z. and D. Szynal (1985), On the expected values of kth records and associated characterizations of distributions, in: *Probability and Statistical Decision Theory*, Vol. A (F. Konecny, J. Mogyoródy, and W. Wertz, eds.) Reidel, Dordrecht, 119–127.

[36] Gumbel, E.J. (1954), The maxima of the mean largest value and of the range, *Ann. Math. Statist.* 25, 76–84.

[37] Hampel, F.R., E.M. Ronchetti, P.J. Rousseeuw, and W.A. Stahel (1986), *Robust Statistics. The Approach Based on Influence Functions*, Wiley, New York.

[38] Hartley, H.O. and H.A. David (1954), Universal bounds for mean range and extreme observation, *Ann. Math. Statist.* 25, 85–99.

[39] Hawkins, D.M. (1971), On the bounds of the range of order statistics, *J. Amer. Statist. Assoc.* 66, 644–645.

[40] Huang, M.L. and P. Brill (1999), A level crossing quantile estimation method, *Statist. Probab. Lett.* 45, 111–119.

[41] Huber, P.J. (1981), *Robust Statistics*, Wiley, New York.

[42] Kamps, U. (1995), *A Concept of Generalized Order Statistics*, Teubner, Stuttgart.

[43] Karlin, S. (1957), P olya type distributions, II, *Ann. Math. Statist.* 28, 281–308.

[44] Karlin, S. (1968), *Total Positivity*, Vol. I, Stanford Univ. Press, Stanford, CA.

[45] Karlin, S. and Studden, W.J. (1966), *Tchebyshev Systems: With Applications in Analysis and Statistics*, Interscience, New York.

[46] Klefsjö, B. (1983), A useful ageing property based on the Laplace transform, *J. Appl. Prob.* 20, 615–626.

[47] Lai, T.L. and H. Robbins (1976), Maximally dependent random variables, *Proc. Nat. Acad. Sci. U.S.A.* 73, 286–288.

[48] Lai, T.L. and H. Robbins (1978), A class of dependent random variables and their maxima, *Z. Wahrsch. Verw. Gebiete* 42, 89–111.

[49] Lawrence, M.J. (1975), Inequalities for s-ordered distributions, *Ann. Statist.* 3, 413–428.

[50] López-Blázquez, F. (1998), Discrete distributions with maximum value of the maximum, *J. Statist. Plann. Inference* 70, 201–207.

[51] López-Blázquez, F. (2000), Bounds for the expected value of spacings from discrete distributions, *J. Statist. Plann. Inference* 84, 1–9.

[52] Ma, C. (1992), Variance bound of function of order statistic, *Statist. Probab. Lett.* 13, 25–27.

[53] Mallows, C.L. (1969), Extrema of expectations of uniform order statistics, *SIAM Rev.* 11, 410–411.

[54] Mallows, C.L. and D. Richter (1969), Inequalities of Chebyshev type involving conditional expectations, *Ann. Math. Statist.* 40, 1922–1932.

[55] Marshall, A.W. and I. Olkin (1979), *Inequalities: Theory of Majorization and Its Applications*, Academic, New York.

[56] Marshall, A.W. and F. Proschan (1970), Mean life of series and parallel systems, *J. Appl. Probab.* 7, 165–174.

[57] Moriguti, S. (1951), Extremal properties of extreme value distributions, *Ann. Math. Statist.* 22, 523–536.

[58] Moriguti, S. (1953), A modification of Schwarz's inequality with applications to distributions, *Ann. Math. Statist.* 24, 107–113.

[59] Nagaraja, H.N. (1978), On the expected values of records, *Austral. J. Statist.*, 20, 176–182.

[60] Nagaraja, H.N. (1981), Some finite sample results for the selection differential, *Ann. Inst. Statist. Math.* 33, 437–448.

[61] Nagaraja, H.N. (1988), Some characterizations of continuous distributions based on regressions of adjacent order statistics and record values, *Sankhyā A* 50, 70–73.

[62] Nagaraja, H.N. (1992), Order statistics from discrete distributions (with discussion), *Statistics* 23, 189–216.

[63] Nair, K.R. (1948), The distribution of the extreme deviate from the sample mean and its studentized form, *Biometrika* 35, 118–144.

[64] Nevzorov, V.B. and N. Balakrishnan (1998), A record of records, in: *Order Statistics: Theory & Methods* (N. Balakrishnan and C.R. Rao, eds.), Handbook of Statistics, Vol. 16, North Holland, Amsterdam, 515–570.

[65] Okolewski, A. and T. Rychlik (2001), Sharp distribution-free bounds on the bias in estimating quantiles via order statistics, *Statist. Probab. Lett.*, to appear.

[66] Olkin, I. (1992), A matrix formulation on how deviant can an observation be, *Amer. Statist.* 46, 205–209.

[67] Papadatos, N. (1995), Maximum variance of order statistics, *Ann. Inst. Statist. Math.* 47, 185–193.

[68] Papadatos, N. (1997), A note on maximum variance of order statistics from symmetric populations, *Ann. Inst. Statist. Math.* 49, 117–121.

[69] Papadatos, N. (1997), Exact bounds for the expectations of order statistics from non-negative populations, *Ann. Inst. Statist. Math.* 49, 727–736.

[70] Papadatos, N. (1999), Upper bound for the covariance of extreme order statistics from a sample of size three, *Sankhyā A* 61, 229–240.

[71] Papadatos, N. (2001), Distribution and expectation bounds on order statistics from possibly dependent variates, *Statist. Probab. Lett.*, to appear.

[72] Plackett, R.L. (1947), Limits of the ratio of mean range to standard deviation, *Biometrika* 34, 120–122.

[73] Prakasa Rao, B.L.S. (1983), *Nonparametric Functional Estimation*, Academic, Orlando, FL.

[74] Raqab, M.Z. (1997), Bounds based on greatest convex minorants for moments of record values, *Statist. Probab. Lett.* 36, 35–41.

[75] Raqab, M.Z. (2000), On the moments of record values, *Commun. Statist. Theory Meth.* 29, 1631–1647.

[76] Rychlik, T. (1992), Stochastically extremal distributions of order statistics for dependent samples, *Statist. Probab. Lett.* 13, 337–341.

[77] Rychlik, T. (1992), Weak limit theorems for stochastically largest order statistics, in: *Order Statistics and Nonparametrics. Theory and Applications* (I.A. Salama and P.K. Sen, eds.), North-Holland, Amsterdam, 141–154.

[78] Rychlik, T. (1992), Sharp inequalities for linear combinations of elements of monotone sequences, *Bull. Polish Acad. Sci. Math.* 40, 247–254.

[79] Rychlik, T. (1993), Bounds for expectation of L-estimates for dependent samples, *Statistics* 24, 1–7.

[80] Rychlik, T. (1993), Bias-robustness of L-estimates of location against dependence, *Statistics* 24, 9–15.

[81] Rychlik, T. (1993), Sharp bounds on L-estimates and their expectations for dependent samples, *Commun. Statist. Theory Meth.* 22, 1053–1068. Erratum in *Commun. Statist. Theory Meth.* 23, 305–306.

[82] Rychlik, T. (1995), Bounds for order statistics based on dependent variables with given nonidentical distributions, *Statist. Probab. Lett.* 23, 351–358.

[83] Rychlik T. (1997), Evaluating improvements of records, *Appl. Math. (Warsaw)* 24, 315–324.

[84] Rychlik T. (1998), Bounds on expectations of L-estimates, in: *Order Statistics: Theory & Methods* (N. Balakrishnan and C.R. Rao, eds.), Handbook of Statistics, Vol. 16, North-Holland, Amsterdam, 105–145.

[85] Rychlik, T. (1999), Error reduction in density estimation under shape restrictions, *Canad. J. Statist.* 27, 607–622.

[86] Rychlik, T. (2000), Evaluating statistical functionals by means of projections onto convex cones in Hilbert spaces: Part I and II, in: *Applied Mathematics Reviews* Vol. 1 (G.A. Anastassiou, ed.) World Sci., Singapore, 407–489.

[87] Rychlik, T. (2001), Mean-variance bounds for order statistics from dependent DFR, IFR, DFRA and IFRA samples, *J. Statist. Plann. Inference*, to appear.

[88] Rychlik, T. (2001), Stability of order statistics under dependence, *Ann. Inst. Statist. Math.*, to appear.

[89] Rychlik, T. (2001), Optimal mean-variance bounds on order statistics from families determined by star order, submitted for publication.

[90] Rychlik, T. (2001), Sharp mean-variance inequalities for quantiles of distributions determined by convex and star orders, submitted for publication.

[91] Rychlik, T. (2001), Predictions of increments of order and record statistics in nonparametric families of distributions, in preparation.

[92] Samuelson, P.A. (1968), How deviant can you be? *J. Amer. Statist. Assoc.* 63, 1522–1525.

[93] Schoenberg, I.J. (1959), On variation diminishing approximation methods, in: On Numerical Approximation: Proc. of Symp., Madison, 1958 (R.E. Langer, ed.), Univ. Wisconsin Press, Madison.

[94] Scott, J.M.C. (1936), Appendix to paper by Pearson and Chandra Sekar, *Biometrika* 28, 319–320.

[95] Serfling, R.J. (1980), *Approximation Theorems of Mathematical Statistics*, Wiley, New York.

[96] Shaked, M. and J.G. Shantikumar (1994), *Stochastic Orders and Their Applications*, Academic, Boston.

[97] Sheather, S.J. and J.S. Marron (1990), Kernel quantile estimators, *J. Amer. Statist. Assoc.* 80, 410–416.

[98] Tchen, A. (1980), Inequalities for distributions with given marginals, *Ann. Probab.* 8, 814–827.

[99] Thompson, W.R. (1935), On a criterion for the rejection of observations and the distribution of the ratio of deviation to sample standard deviation, *Ann. Math. Statist.* 6, 214–219.

[100] van Zwet, W.R. (1964), *Convex Transformations of Random Variables*, Math. Centre Tracts, Vol. 7, Mathematisch Centrum, Amsterdam.

[101] von Mises, M. (1947), On the asymptotic distribution of differentiable statistical functions, *Ann. Math. Statist* 18, 309–348.

[102] Vysochanskii, D.F. and Y. Petunin (1979), Justification of the three-sigma rule for unimodal distributions, *Theor. Probab. Math. Statist.* 21, 25–36.

[103] Zieliński, R. (1988), A distribution-free median-unbiased quantile estimator, *Statistics* 19, 223–227.

[104] Zieliński, R. (1998), Uniform strong consistency of sample quantiles, *Statis. Probab. Lett.* 37, 115–119.

[105] Zieliński, R. (1999), Best equivariant nonparametric estimator of a quantile, *Statist. Probab. Lett.* 45, 79–84.

[106] Zieliński, R. (2001), PMC-optimal nonparametric quantile estimator, *Statistics*, to appear.

Author Index

Ahsanullah, M., 20, 163
Anastassiou, G.A., 169
Arnold, B.C., 18, 20, 24, 55, 101,
    103, 104, 131, 133, 163

Balakrishnan, A. V., 11, 12, 163
Balakrishnan, N., 18, 24, 55, 58,
    96, 104, 131, 133, 163,
    164, 167, 169
Barlow, R.E., 28, 164
Biondini, R., 133, 164
Blom, G., 55, 164
Boyd, A.V., 104, 164
Brill, P., 82, 166

Caraux, G., 19, 99, 101, 103, 164,
    165
Chandler, K.N., 20, 164
Charalambides, C., 164
Cheng, C., 82, 164
Cohen, A.C., 18, 164

David, H.A., 2, 18, 55, 56, 104,
    164–166
Dharmadhikari, S., 27–29, 33, 165

Dziubdziela, W., 21, 165

Fahmy, S., 104, 165
Feldman, D., 83, 165
Franco, M., 24, 25, 165

Gajek, L., v, 3, 55, 59, 68, 81, 96,
    118, 128, 131, 137, 165
Gascuel, O., 19, 99, 101, 103, 164,
    165
Groeneveld, R.A., 104, 163
Grudzień, Z., 135, 165
Gumbel, E.J., 2, 56, 165

Hampel, F.R., 123, 165
Hartley, H.O., 2, 56, 165, 166
Hawkins, D.M., 104, 166
Huang, M.L., 82, 166
Huber, P.J., 123, 166

Joag-dev, K., 27–29, 33, 165

Kamps, U., 158, 166
Karlin, S., 66, 84, 138, 166
Klefsjö, B., 54, 166
Konecny, F., 165

Kopociński, B., 21, 165

Lai, T.L., 99, 166
Langer, R.E., 169
Lawrence, M.J., 28, 166
López-Blázquez, F., 55, 166

Ma, C., 55, 166
Mallows, C.L., 99, 104, 166
Marron, J.S., 82, 169
Marshall, A.W., 14, 15, 167
Mogyoródy, J., 165
Moriguti, S., 3, 16, 33, 35, 55, 56, 59, 93, 118, 133, 141, 148, 159, 167

Nagaraja, H.N., 24, 25, 56, 131, 133, 163, 167
Nair, K.R., 104, 167
Nevzorov, V.B., 24, 167

Okolewski, A., 55, 84, 131, 137, 165, 167
Olkin, I., 14, 104, 167

Papadatos, N., 55, 96, 164, 167, 168
Pearson, E.S., 165
Petunin, Y., 33, 170
Plackett, R.L., 56, 93, 168
Prakasa Rao, B.L.S., 16, 168
Proschan, F., 15, 28, 104, 164, 165, 167
Puri, M.L., 164

Rao, C.R., 18, 164, 167, 169
Raqab, M.Z., 131, 133, 135, 168
Richter, D., 104, 166
Robbins, H., 99, 166
Ronchetti, E.M., 165
Rousseeuw, P.J., 165
Ruiz, J.M., 24, 25, 165
Rychlik, T., v, 3, 16, 18, 19, 30, 33, 55, 56, 59, 68, 78, 81, 84, 95, 96, 99, 103, 104, 111, 118, 123, 127, 128, 131, 146, 165, 167–169

Salama, I.A., 168
Samuelson, P.A., 104, 169
Schoenberg, I.J., 65, 169
Scott, J.M.C., 104, 169
Sen, P.K., 168
Serfling, R.J., 16, 169
Shaked, M., 29, 48, 169
Shantikumar, J.G., 29, 48, 169
Sheather, S.J., 82, 169
Siddiqui, M.M., 133, 164
Stahel, W.A., 165
Studden, W.J., 66, 138, 166
Szynal, D., 135, 165

Tchen, A., 99, 169
Thompson, W.R., 104, 170
Tucker, H.G., 83, 165

van Zwet, W.R., 27, 28, 55, 170
von Mises, M., 16, 170
Vysochanskii, D.F., 33, 170

Wertz, W., 165

Zieliński, R., 82, 83, 170

Subject Index

atom, see degenerate distribution

Bernstein polynomial, 65, 66, 82, 137
bias, 82–92, 118–122
binomial distribution, 58

Chebyshev inequality, 33
Chebyshev system, 66
conditional expectation
    of order statistics, 22–24, 146–148, 150–153
    of record values, 24, 25, 148, 149, 153, 154
convergence
    almost sure, 82, 91, 93
    in mean square, 31
    weak, 31
convex cone, 1, 2, 12–16, 25–31
convex order, 27, 28, 36–44, 60–68, 84–89, 105–114, 119, 120, 136–140, 149–154

de Moivre–Laplace theorem, 83
decreasing density (DD) distribution, 27, 40, 53, 66, 67, 89, 92, 106, 111, 117, 120, 122, 138, 139, 142, 151–153

decreasing density on the average (DDA) distribution, 29, 48, 53, 76, 77, 81, 90, 92, 115, 117, 120, 122

decreasing failure rate (DFR) distribution, 28, 40, 41, 53, 67, 68, 89, 92, 106, 111, 117, 120, 122, 139, 140, 142, 152–154

decreasing failure rate on the average (DFRA) distribution, 28, 29, 48, 53, 77, 78, 81, 91, 92, 115, 117, 121, 122

degenerate distribution, 40, 41, 44, 48, 58, 60, 89, 106, 107, 109, 111, 113, 116, 148, 149, 151, 153, 154

Dirac distribution, see degenerate distribution

Subject Index

exponential distribution, 20, 40, 41, 44, 48, 67, 89, 106, 109, 111, 114, 139, 142, 153

failure rate, 28

gamma distribution, 20, 137, 140
Gauss inequality, 33
general distribution, 25, 26, 34, 53, 57, 81, 82, 92, 100, 101, 117–119, 122, 125, 126, 134, 141, 147–149
generalized order statistic, 158
Gini mean difference, 93
greatest convex minorant, 3, 14, 34, 35, 57, 84, 93, 118, 123, 124, 133, 135, 141, 143, 147, 159, 160

hazard function, 28, 159
hazard rate, see failure rate
Hilbert space, 1, 2, 11–16
Hölder inequality, 103, 131, 159, 160

increasing density (ID) distribution, 27, 44, 53, 109, 113, 117, 142
increasing density on the average (IDA) distribution, 29, 48, 79, 115
increasing failure rate (IFR) distribution, 28, 44, 53, 109, 113, 114, 117, 142
increasing failure rate on the average (IFRA) distribution, 28, 29, 48, 79, 115
inner product, 2, 11, 13
inner product space, 11

j-out-of-n reliability system, 18, 60
Jensen inequality, 55, 131
jump, see degenerate distribution

kth lower record value, 22, 143
kth (upper) record occurrence time, 21
kth (upper) record value, 21, 22, 131–143, 148, 149, 153, 154

L-statistic, 18, 19, 56, 93–97, 100–104, 115, 116, 121, 123, 128, 129, 157, 158
Laplace transform order, 48
Lebesgue monotone convergence theorem, 15
life distribution, 26, 58, 60–68, 78, 107, 136–140
linear functional, 1, 11, 16–25

Markov inequality, 33
Markov property, 24, 133

negative binomial distribution, 58
new better than used (NBU) distribution, 54, 159
new worse than used (NWU) distribution, 159
norm, 1, 11

order statistic
    of dependent sample, 18, 19, 95–129
    of independent sample, 2, 3, 17, 18, 55–93, 123–127, 146–148, 150–153

Poisson distribution, 159
power distribution, 2, 56, 148, 149
prediction
    of order statistics, 146–148, 150–153
    of records, 25, 148, 149, 153, 154
projection, 1–3, 12–16, 29–31

quantile, 17, 33–54, 82–92, 118–122
quantile function, 2, 16, 17, 25–31

record increment, 140–142, 148, 149, 153, 154
record occurence time, 20
record value, 20, 21, 131–143, 148, 149
Riesz representation theorem, 11
robust estimate, 19, 93, 121, 123, 127

s-order, 28, 50–52, 79–81, 115, 116
sample interquartile distance, 18
sample maximum, 55, 56, 58, 59, 99, 101, 103, 104, 126, 127
sample median, 18
sample minimum, 57, 58, 125, 127
sample quantile, 82, 118
sample quasirange, 93
sample range, 18, 55, 56, 93, 103, 104, 129
Schwarz inequality, 2, 11, 13, 46, 59, 68, 74, 87, 93, 101–104, 133, 135, 136, 146, 151, 159
selection differential, 56, 104
spacing, 93, 128, 147, 150
star order, 28, 29, 44–50, 69–79, 89–91, 114, 115, 120, 121
starshaped function, 28
superadditive function, 158
superadditive order, 48, 158
symmetric distribution, 26, 35, 53, 60, 81, 102, 117
symmetric U-shaped distribution, 28, 116, 117
symmetric unimodal distribution, 28, 52, 53, 79–81, 115–117

three-point distribution, 35, 102, 119
totally positive function, 84, 137
trimmed mean, 18, 93, 128
two-point distribution, 34, 58, 82, 101, 116, 119, 126, 147

uniform distribution, 27, 40, 44, 48, 66, 79, 89, 106, 109, 111, 113, 115, 116, 138, 142, 151, 153

variation diminishing property, 65, 137, 138

Weibull distribution, 133, 149
Winsorized mean, 128

Lecture Notes in Statistics

For information about Volumes 1 to 86, please contact Springer-Verlag

Vol. 87: J. Müller, Lectures on Random Voronoi Tessellations. vii, 134 pages, 1994.

Vol. 88: J. E. Kolassa, Series Approximation Methods in Statistics. Second Edition. ix, 183 pages, 1997.

Vol. 89: P. Cheeseman, and R.W. Oldford (Editors), Selecting Models from Data: AI and Statistics IV. xii, 487 pages, 1994.

Vol. 90: A. Csenki, Dependability for Systems with a Partitioned State Space: Markov and Semi-Markov Theory and Computational Implementation. x, 241 pages, 1994.

Vol. 91: J.D. Malley, Statistical Applications of Jordan Algebras. viii, 101 pages, 1994.

Vol. 92: M. Eerola, Probabilistic Causality in Longitudinal Studies. vii, 133 pages, 1994.

Vol. 93: Bernard Van Cutsem (Editor), Classification and Dissimilarity Analysis. xiv, 238 pages, 1994.

Vol. 94: Jane F. Gentleman and G.A. Whitmore (Editors), Case Studies in Data Analysis. viii, 262 pages, 1994.

Vol. 95: Shelemyahu Zacks, Stochastic Visibility in Random Fields. x, 175 pages, 1994.

Vol. 96: Ibrahim Rahimov, Random Sums and Branching Stochastic Processes. viii, 195 pages, 1995.

Vol. 97: R. Szekli, Stochastic Ordering and Dependence in Applied Probability. viii, 194 pages, 1995.

Vol. 98: Philippe Barbe and Patrice Bertail, The Weighted Bootstrap. viii, 230 pages, 1995.

Vol. 99: C.C. Heyde (Editor), Branching Processes: Proceedings of the First World Congress. viii, 185 pages, 1995.

Vol. 100: Wlodzimierz Bryc, The Normal Distribution: Characterizations with Applications. viii, 139 pages, 1995.

Vol. 101: H.H. Andersen, M.Højbjerre, D. Sørensen, and P.S. Eriksen, Linear and Graphical Models for the Multivariate Complex Normal Distribution. x, 184 pages, 1995.

Vol. 102: A.M. Mathai, Serge B. Provost, and Takesi Hayakawa, Bilinear Forms and Zonal Polynomials. x, 378 pages, 1995.

Vol. 103: Anestis Antoniadis and Georges Oppenheim (Editors), Wavelets and Statistics. vi, 411 pages, 1995.

Vol. 104: Gilg U.H. Seeber, Brian J. Francis, Reinhold Hatzinger, and Gabriele Steckel-Berger (Editors), Statistical Modelling: 10th International Workshop, Innsbruck, July 10–14th, 1995. x, 327 pages, 1995.

Vol. 105: Constantine Gatsonis, James S. Hodges, Robert E. Kass, and Nozer D. Singpurwalla (Editors), Case Studies in Bayesian Statistics, Volume II. x, 354 pages, 1995.

Vol. 106: Harald Niederreiter and Peter Jau-Shyong Shiue (Editors), Monte Carlo and Quasi-Monte Carlo Methods in Scientific Computing. xiv, 372 pages, 1995.

Vol. 107: Masafumi Akahira and Kei Takeuchi, Non-Regular Statistical Estimation. vii, 183 pages, 1995.

Vol. 108: Wesley L. Schaible (Editor), Indirect Estimators in US Federal Programs. viii, 195 pages, 1995.

Vol. 109: Helmut Rieder (Editor), Robust Statistics, Data Analysis, and Computer Intensive Methods. xiv, 427 pages, 1996.

Vol. 110: D. Bosq, Nonparametric Statistics for Stochastic Processes. xii, 169 pages, 1996.

Vol. 111: Leon Willenborg and Ton de Waal, Statistical Disclosure Control in Practice. xiv, 152 pages, 1996.

Vol. 112: Doug Fischer and Hans-J. Lenz (Editors), Learning from Data. xii, 450 pages, 1996.

Vol. 113: Rainer Schwabe, Optimum Designs for Multi-Factor Models. viii, 124 pages, 1996.

Vol. 114: C.C. Heyde, Yu. V. Prohorov, R. Pyke, and S.T. Rachev (Editors), Athens Conference on Applied Probability and Time Series Analysis Volume I: Applied Probability in Honor of J.M. Gani. viii, 424 pages, 1996.

Vol. 115: P.M. Robinson and M. Rosenblatt (Editors), Athens Conference on Applied Probability and Time Series Analysis Volume II: Time Series Analysis in Memory of E.J. Hannan. viii, 448 pages, 1996.

Vol. 116: Genshiro Kitagawa and Will Gersch, Smoothness Priors Analysis of Time Series. x, 261 pages, 1996.

Vol. 117: Paul Glasserman, Karl Sigman, and David D. Yao (Editors), Stochastic Networks. xii, 298, 1996.

Vol. 118: Radford M. Neal, Bayesian Learning for Neural Networks. xv, 183, 1996.

Vol. 119: Masanao Aoki and Arthur M. Havenner, Applications of Computer Aided Time Series Modeling. ix, 329 pages, 1997.

Vol. 120: Maia Berkane, Latent Variable Modeling and Applications to Causality. vi, 288 pages, 1997.

Vol. 121: Constantine Gatsonis, James S. Hodges, Robert E. Kass, Robert McCulloch, Peter Rossi, and Nozer D. Singpurwalla (Editors), Case Studies in Bayesian Statistics, Volume III. xvi, 487 pages, 1997.

Vol. 122: Timothy G. Gregoire, David R. Brillinger, Peter J. Diggle, Estelle Russek-Cohen, William G. Warren, and Russell D. Wolfinger (Editors), Modeling Longitudinal and Spatially Correlated Data. x, 402 pages, 1997.

Vol. 123: D.Y. Lin and T.R. Fleming (Editors), Proceedings of the First Seattle Symposium in Biostatistics: Survival Analysis. xiii, 308 pages, 1997.

Vol. 124: Christine H. Müller, Robust Planning and Analysis of Experiments. x, 234 pages, 1997.

Vol. 125: Valerii V. Fedorov and Peter Hackl, Model-Oriented Design of Experiments. viii, 117 pages, 1997.

Vol. 126: Geert Verbeke and Geert Molenberghs, Linear Mixed Models in Practice: A SAS-Oriented Approach. xiii, 306 pages, 1997.

Vol. 127: Harald Niederreiter, Peter Hellekalek, Gerhard Larcher, and Peter Zinterhof (Editors), Monte Carlo and Quasi-Monte Carlo Methods. xii, 448 pages, 1997.

Vol. 128: L. Accardi and C.C. Heyde (Editors), Probability Towards 2000, x, 356 pages, 1998.

Vol. 129: Wolfgang Härdle, Gerard Kerkyacharian, Dominique Picard, and Alexander Tsybakov, Wavelets, Approximation, and Statistical Applications. xvi, 265 pages, 1998.

Vol. 130: Bo-Cheng Wei, Exponential Family Nonlinear Models. ix, 240 pages, 1998.

Vol. 131: Joel L. Horowitz, Semiparametric Methods in Econometrics. ix, 204 pages, 1998.

Vol. 132: Douglas Nychka, Walter W. Piegorsch, and Lawrence H. Cox (Editors), Case Studies in Environmental Statistics. viii, 200 pages, 1998.

Vol. 133: Dipak Dey, Peter Müller, and Debajyoti Sinha (Editors), Practical Nonparametric and Semiparametric Bayesian Statistics. xv, 408 pages, 1998.

Vol. 134: Yu. A. Kutoyants, Statistical Inference for Spatial Poisson Processes. vii, 284 pages, 1998.

Vol. 135: Christian P. Robert, Discretization and MCMC Convergence Assessment. x, 192 pages, 1998.

Vol. 136: Gregory C. Reinsel and Raja P. Velu, Multivariate Reduced-Rank Regression. xiii, 272 pages, 1998.

Vol. 137: V. Seshadri, The Inverse Gaussian Distribution: Statistical Theory and Applications. xi, 360 pages, 1998.

Vol. 138: Peter Hellekalek and Gerhard Larcher (Editors), Random and Quasi-Random Point Sets. xi, 352 pages, 1998.

Vol. 139: Roger B. Nelsen, An Introduction to Copulas. xi, 232 pages, 1999.

Vol. 140: Constantine Gatsonis, Robert E. Kass, Bradley Carlin, Alicia Carriquiry, Andrew Gelman, Isabella Verdinelli, and Mike West (Editors), Case Studies in Bayesian Statistics, Volume IV. xvi, 456 pages, 1999.

Vol. 141: Peter Müller and Brani Vidakovic (Editors), Bayesian Inference in Wavelet Based Models. xi, 394 pages, 1999.

Vol. 142: György Terdik, Bilinear Stochastic Models and Related Problems of Nonlinear Time Series Analysis: A Frequency Domain Approach. xi, 258 pages, 1999.

Vol. 143: Russell Barton, Graphical Methods for the Design of Experiments. x, 208 pages, 1999.

Vol. 144: L. Mark Berliner, Douglas Nychka, and Timothy Hoar (Editors), Case Studies in Statistics and the Atmospheric Sciences. x, 208 pages, 2000.

Vol. 145: James H. Matis and Thomas R. Kiffe, Stochastic Population Models. viii, 220 pages, 2000.

Vol. 146: Wim Schoutens, Stochastic Processes and Orthogonal Polynomials. xiv, 163 pages, 2000.

Vol. 147: Jürgen Franke, Wolfgang Härdle, and Gerhard Stahl, Measuring Risk in Complex Stochastic Systems. xvi, 272 pages, 2000.

Vol. 148: S.E. Ahmed and Nancy Reid, Empirical Bayes and Likelihood Inference. x, 200 pages, 2000.

Vol. 149: D. Bosq, Linear Processes in Function Spaces: Theory and Applications. xv, 296 pages, 2000.

Vol. 150: Tadeusz Caliński and Sanpei Kageyama, Block Designs: A Randomization Approach, Volume I: Analysis. ix, 313 pages, 2000.

Vol. 151: Håkan Andersson and Tom Britton, Stochastic Epidemic Models and Their Statistical Analysis. ix, 152 pages, 2000.

Vol. 152: David Ríos Insua and Fabrizio Ruggeri, Robust Bayesian Analysis. xiii, 435 pages, 2000.

Vol. 153: Parimal Mukhopadhyay, Topics in Survey Sampling. x, 303 pages, 2000.

Vol. 154: Regina Kaiser and Agustín Maravall, Measuring Business Cycles in Economic Time Series. vi, 190 pages, 2000.

Vol. 155: Leon Willenborg and Ton de Waal, Elements of Statistical Disclosure Control. xvii, 289 pages, 2000.

Vol. 156: Gordon Willmot and X. Sheldon Lin, Lundberg Approximations for Compound Distributions with Insurance Applications. xi, 272 pages, 2000.

Vol. 157: Anne Boomsma, Marijtje A.J. van Duijn, and Tom A.B. Snijders (Editors), Essays on Item Response Theory. xv, 448 pages, 2000.

Vol. 158: Dominique Ladiray and Benoît Quenneville, Seasonal Adjustment with the X-11 Method. xxii, 225 pages, 2001.

Vol. 159: Marc Moore (Editor), Spatial Statistics: Methodological Aspects and Some Applications. xvi, 282 pages, 2001.

Vol. 160: Tomasz Rychlik, Projecting Statistical Functionals. ix, 169 pages, 2001.